Elements of
CELL BIOLOGY

Elements of
CELL BIOLOGY

G. Jahir Hussain

ANMOL PUBLICATIONS PVT. LTD.
NEW DELHI-110 002 (INDIA)

ANMOL PUBLICATIONS PVT. LTD.

Regd. Office: 4360/4, Ansari Road, Daryaganj,
New Delhi-110002 (India)
Tel.: 23278000, 23261597, 23286875, 23255577
Fax: 91-11-23280289
Email: anmolpub@gmail.com
Visit us at: www.anmolpublications.com

Branch Office: No. 1015, Ist Main Road, BSK IIIrd Stage
IIIrd Phase, IIIrd Block, Bengaluru-560 085 (India)
Tel.: 080-41723429 • Fax: 080-26723604
Email: anmolpublicationsbangalore@gmail.com

Elements of Cell Biology

First Edition, 2011

PRINTED IN INDIA

Printed at Mehra Offset Press, Delhi.

Contents

Preface

Cell biology has become the vital aspects in modern life science. As it is a base for Molecular and other emerging sciences, it is contributing significantly in understanding the molecular basis of life. It basically explains the molecular and anatomical level of cell organization.

The structure and composition of living organisms varies greatly from single celled bacteria to complex multi-cellular organisms with differentiated cell types and interconnected organ systems. Regardless of the complexity, every living entity contains a blueprint for its construction in the form of a double helical chain of molecules called Deoxyribonucleic Acid (DNA) the genetic material. The DNA is an amazingly simple chemical structure, yet it contains an entire library of information on how to make, maintain, and reproduce and organism, and also keeps a record of clues to the organism's evolutionary history.

Cell biology research encompasses both the great diversity of single-celled organisms like bacteria and protozoa, as well as the many specialized cells in multicellular organisms such as humans. Knowing the components of cells and how cells work is fundamental to all biological sciences. Appreciating the similarities and differences between cell types is particularly important to the fields of cell and molecular biology as well as to biomedical fields such as cancer research and developmental biology.

Author

1

Introduction

Cell biology is a scientific discipline that studies cells–their physiological properties, their structure, the organelles they contain, interactions with their environment, their life cycle, division and death. This is done both on a microscopic and molecular level. Cell biology research encompasses both the great diversity of single-celled organisms like bacteria and protozoa, as well as the many specialized cells in multicellular organisms such as humans.

Knowing the components of cells and how cells work is fundamental to all biological sciences. Appreciating the similarities and differences between cell types is particularly important to the fields of cell and molecular biology as well as to biomedical fields such as cancer research and developmental biology.

These fundamental similarities and differences provide a unifying theme, sometimes allowing the principles learned from studying one cell type to be extrapolated and generalized to other cell types. Hence, research in cell biology is closely related to genetics, biochemistry, molecular biology, immunology, and developmental biology.

The cell is the functional basic unit of life. It was discovered by Robert Hooke and is the functional unit of all known living organisms. It is the smallest unit of life that is classified as a living thing, and is often called the building block of life. Some organisms, such as most bacteria, are unicellular (consist of a single cell). Other organisms, such as humans, are multicellular. Humans have about 100 trillion or 1014 cells; a typical cell size is 10 μm and a typical cell mass is

1 nanogram. The largest cells are about 135 μm in the anterior horn in the spinal cord while granule cells in the cerebellum, the smallest, can be some 4 μm and the longest cell can reach from the toe to the lower brain stem (Pseudounipolar cells). The largest known cells are unfertilised ostrich egg cells which weigh 3.3 pounds.

In 1835, before the final cell theory was developed, Jan Evangelista Purkynì observed small "granules" while looking at the plant tissue through a microscope. The cell theory, first developed in 1839 by Matthias Jakob Schleiden and Theodor Schwann, states that all organisms are composed of one or more cells, that all cells come from preexisting cells, that vital functions of an organism occur within cells, and that all cells contain the hereditary information necessary for regulating cell functions and for transmitting information to the next generation of cells.

THE MOLECULES OF LIFE

Among the many events that occur in the life of a cell are a multitude of specific chemical transformations, which provide the cell with usable energy and the molecules needed to form its structure and coordinate its activities. Here we briefly describe the functions of the main types of chemicals that compose cells.

Water, inorganic ions, and a large array of relatively small organic molecules (*e.g.*, sugars, vitamins, fatty acids) account for 75 – 80 per cent of living matter by weight. Of these small molecules, water is by far the most abundant. The remainder of living matter consists of macromolecules, including proteins, polysaccharides, and DNA.

Cells acquire and use these two size classes of molecules in fundamentally different ways. Ions, water, and many small organic molecules are imported into the cell. Cells also make and alter many small organic molecules by a series of different chemical reactions.

In contrast, cells can obtain macromolecules only by making them. Their synthesis entails linking together a specific set of small molecules (monomers) to form polymers through repetition of a single type of chemical-linkage reaction. Some small molecules function as precursors for synthesis of macromolecules, and the cell is careful to provide the appropriate mix of small molecules needed.

Small molecules also store and distribute the energy for all cellular processes; they are broken down to extract this chemical energy, as when sugar is degraded to carbon dioxide and water with the release of the energy bound up in the molecule.

Other small molecules (*e.g.*, hormones and growth factors) act as signals that direct the activities of cells and nerve cells communicate with one another by releasing and sensing certain small signaling molecules. The powerful effect on our body of a frightening event comes from the instantaneous flooding of the body with a small-molecule hormone that mobilizes the "fight or flight" response.

Macromolecules, though, are the most interesting and characteristic molecules of living systems; in a true sense the evolution of life as we know it is the evolution of macromolecular structures. Proteins, the workhorses of the cell, are the most abundant and functionally versatile of the cellular macromolecules.

To appreciate the abundance of protein within a cell, we can estimate the number of protein molecules in a typical eukaryotic cell, such as a hepatocyte in the liver. This cell, roughly a cube 15 μm (0.0015 cm) on a side, has a volume of 3.4×10^{-9} cm^3 (or milliliters).

Assuming a cell density of 1.03 g/ml, the cell would weigh 3.5×10^{-9} g. Since protein accounts for approximately 20 per cent of a cell's weight, the total weight of cellular protein is 7×10^{-10} g. The average yeast protein has a molecular weight of 52,700 (g/mol). Assuming this value is typical of eukaryotic proteins, we can calculate the total number of protein molecules per liver cell as about 7.9×10^9 from the total protein weight and the number of molecules per mole, which is a constant (Avogadro's number).

To carry this calculation one step further, consider that a liver cell contains about 10,000 different proteins; thus, a cell contains close to a million molecules of each protein on average. In actuality, however, the abundance of different proteins varies widely, from the quite rare cell-surface protein that binds the hormone insulin (20,000 molecules) to the abundant structural protein actin (5×10^8 molecules).

Many of the proteins within cells are enzymes, which accelerate (catalyze) reactions involving small molecules. Other proteins allow cells to move and do work, maintain internal cell rigidity, and transport molecules across membranes. Proteins even direct their own synthesis and that of other macromolecules. Reflecting their numerous functions, proteins come in many shapes and sizes.

The elucidation of the structure of proteins and the relation of protein structure to function remain active areas of scientific investigation. Proteins are formed from only 20 different monomers, the amino acids. That such a limited set of building blocks can do so much is a continuous marvel, even to researchers who work with proteins every day. They are the true glory of the biological world.

The macromolecule that garners the most public attention is not protein but deoxyribonucleic acid (DNA), whose functional properties make it the cell's master molecule. The three-dimensional structure of DNA, first proposed by James D. Watson and Francis H. C. Crick about 50 years ago, consists of two long helical strands that are coiled around a common axis forming a double helix.

The double-helical structure of DNA, one of nature's most magnificent constructions, is critical to the phenomenon of heredity, the transfer of genetically determined characteristics from one generation to the next.

Each strand of DNA is composed of just four different types of monomers called nucleotides. Genes are simply coded representations of the structures of individual proteins, a code written in four chemical "letters"—the nucleotides—and displayed as a continually varying sequence in DNA. Since cells use proteins (enzymes) to make other molecules like sugars or fats, DNA indirectly directs the synthesis of many small molecules as well as proteins. DNA also contains a coded set of instructions about when various proteins are to be made and in what quantities.

In the common view, DNA is the storage form of genetic information, which protein "machines" read out for use by the cell. But a third macromolecule, ribonucleic acid (RNA), is necessary in the process.

The *central dogma* of biology states that the coded genetic information hard-wired into DNA is transcribed into individual transportable cassettes, composed of messenger RNA (mRNA); each mRNA cassette contains the programme for synthesis of a particular protein (or small number of proteins). This critical trio of macromolecules—DNA, RNA, and proteins—is present in all cells.

The mechanism whereby the information encoded in DNA is deciphered into proteins is now understood quite well and explained.

How this process of gene expression is regulated—that is, how cells "know" to make the right proteins at the right time in the right amounts — is a major focus of current research in molecular cell biology and a recurring theme throughout this book.

THE ARCHITECTURE OF CELLS

Although generalizations in biology usually lack the theoretical underpinnings found in physics, there are very clear commonalities among living systems that give biology a unity.

One is the style of cellular construction. The biological universe consists of two types of cells— *prokaryotic cells,* which lack a defined nucleus and have a simplified internal organization, and *eukaryotic cells,* which have a more complicated internal structure including a defined, membrane-limited nucleus. Detailed analysis of the DNA from a variety of prokaryotic organisms in recent years has revealed two distinct types: bacteria (often called "true" bacteria or eubacteria) and archaea (also called *archaebacteria* or *archaeans*). As we discuss the archaea are in some respects more similar to eukaryotic organisms than to the true bacteria.

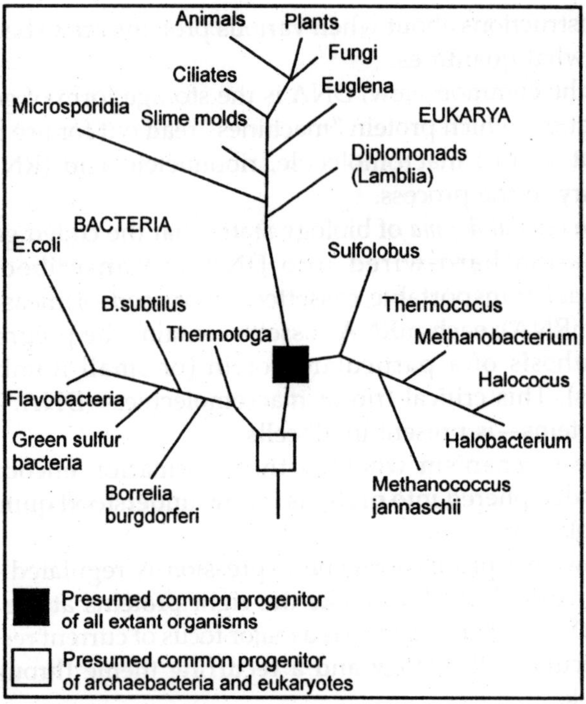

Fig. The three Kingdoms of Organisms are Related through Common Sequences of their Ribosomal RNAs.

Based on the assumption that organisms with more similar genes evolved from a common progenitor more recently than those with more dissimilar genes, researchers have developed the lineage tree. According to this tree, the archaea and eukarya (eukaryotes) are thought to have diverged from the bacteria before they diverged from each other. Despite the differences in the organization of prokaryotic and eukaryotic cells, all cells share certain structural features and carry out many complicated processes in basically the same way.

CELLS ARE SURROUNDED BY WATER- IMPERMEABLE MEMBRANES

A cell, because it is a limited space, must have an outer border. The construction of that border represents one of the most fundamental considerations in biological organization. The outer shell of cells, like any shell, is built to keep the interior contents from leaking out into the surrounding environment. The chemical processes of cellular life generally take place in a watery solution, and the intracellular constituents of cells are largely molecules that are easily dissolved in water. Similarly, the environment around cells is a watery one, the blood and other bodily fluids being solutions in water. Cells then, in order to maintain their integrity, need to be surrounded by an environment through which water cannot flow. A membrane composed of fatty molecules serves this purpose.

We all know from common experience that "oil and water don't mix." That maxim is all one needs to appreciate how a cell is constructed. When oil is poured on water, the oil spreads into a thin film; that film is analogous to the film of fat that surrounds cells, called the plasma membrane. Biological membranes differ from a pure oil film in that the molecules that make the membrane have both oily and watery portions; they have long fatty chains, but they also have a head group that is water-soluble by virtue of being electrically charged.

Thus membranes are formed because these bipartite molecules, called phospholipids, spontaneously orient themselves to form a double layer, or bilayer, having a fatty interior with external surfaces bonded to the surrounding water by the charged head groups.

The membrane is given rigidity by interspersion of cholesterol, a molecule we have come to hate because of its association with heart disease, but one that is required to build the outer membrane of all our cells. Hence from an understanding of the contrasting properties of watery solutions and oily layers, an understanding of cellular construction emerges.

In spite of the rigidity provided by cholesterol, membranes composed of fat are not very strong, so numerous mechanisms for strengthening the borders of cells have evolved. In plants the plasma membrane is surrounded by a rigid cell wall. Although most animal cells lack a cell wall, proteins attached to their exterior surfaces provide some stability; the linking of cells together through these proteins helps maintain the integrity of tissues. Tissues and organs are often covered by strong networks of proteins and other molecules that strengthen and protect them, and also wall off the various compartments of the body. Single-celled organisms, like bacteria, have special outer coats to protect them.

MEMBRANES SERVE FUNCTIONS

Although membranes are valuable as a way to segregate the watery interior of the cell from its environment, or to segregate intracellular events from one another, they have other important functions, including energy storage. Because membranes separate watery compartments from one another, if an ion or a molecule dissolved in water is moved through a membrane into a new cellular compartment, it will not be able to diffuse freely out of the compartment into which it was moved. It takes energy to move the molecule, but once moved, the molecule stores that energy by virtue of its entrapment. Formally, this storage of energy is just like the storage of energy in a battery. Therefore, membranes not only delineate compartments, but also serve as active participants in the cell's dynamism.

The functions of many proteins depend on their mode of association with membranes. For instance, the passage of water-soluble molecules through membranes is carried out by protein transporters that are embedded in the membrane. Also, cells send information to one another by releasing signaling molecules. The outer membranes of cells have proteins, known appropriately as *receptors,* that bind the circulating signaling molecules. These signaling molecules allow the individual activities of the many cells in the body to be coordinated. The receipt of a signaling molecule by a receptor causes the transient organization of particular types of intracellular proteins, called *signal-transduction proteins,* into an activated complex at the interior face of the cell's outer membrane, from which it directs alterations of events in the cell's cytoplasm or nucleus.

PROKARYOTES

All prokaryotes are single-celled organisms, or protists. The bacterial lineage includes *Escherichia coli*, found in animal intestines and a favourite experimental organism, and the photosynthetic organisms formerly known as *blue-green algae* but better known today as *cyanobacteria*. (Because most prokaryotes studied in laboratories are bacteria, discussions of prokaryotic structure or metabolism throughout this book refer to these organisms, not archaeans, unless noted otherwise.) Many members of the archaeal lineage grow in unusual, often extreme, environments. For instance, the halophiles require high concentrations of salt to survive, and the thermoacidophiles grow in hot (80°C) sulfur springs, where a pH of less than 2 is common. Other archaeans, called *methanogens*, live in oxygen-free milieus and generate methane (CH_4) by the reduction of carbon dioxide.

Archaeal cells have a similar structure. In general, prokaryotes consist of a single closed compartment containing the cytosol and bounded by thc plasma membrane. Although bacterial cells do not have a defined nucleus, the genetic material, DNA, is condensed into the central region of the cell. In addition, most ribosomes — the cell's protein-synthesizing particles — are found in the DNA-free region of the cell. Some bacteria also have an invagination of the cell membrane, called a *mesosome*, which is associated with synthesis of DNA and secretion of proteins. Thus bacterial cells are not completely devoid of internal organization.

Bacterial cells possess a cell wall, which lies adjacent to the external side of the plasma membrane. The cell wall is composed of layers of peptidoglycan, a complex of proteins and oligosaccharides; it helps protect the cell and maintain its shape. Some bacteria (*e.g.*, *E. coli*) have a thin cell wall and an unusual outer membrane separated from the cell wall by the periplasmic space. Such bacteria are not stained by the Gram technique and thus are classified as gram-negative. Other bacteria (*e.g.*, *Bacillus polymyxa*) that have a thicker cell wall and no outer membrane take the Gram stain and thus are classified as gram-positive.

EUKARYOTIC CELLS

Eukaryotes comprise all members of the plant and animal kingdoms, including the unicellular fungi (*e.g.*, yeasts, mushrooms, molds) and protozoans. Eukaryotic cells, like prokaryotic cells, are surrounded by a plasma membrane. However, unlike prokaryotic cells, most eukaryotic cells also contain extensive internal membranes that enclose specific compartments, the organelles, and separate them from the rest of the cytoplasm, the region of the cell lying outside the nucleus.

Most organelles are surrounded by a single phospholipid membrane, but several, including the nucleus, are enclosed by two membranes. Each type of organelle plays a unique role in the growth and metabolism of the cell, and each contains a collection of specific enzymes that catalyze requisite chemical reactions. The membranes defining these subcellular compartments control their internal ionic composition so that it commonly differs from that of the cytosol (the portion of the cytoplasm outside the organelles) and among the various organelles.

The largest organelle in a eukaryotic cell is generally the nucleus, which houses most of the cellular DNA. In addition to the nucleus, several other organelles are present in nearly all eukaryotic cells: the mitochondria, in which much of the cell's energy metabolism is carried out; the rough and smooth endoplasmic reticula, a network of membranes in which glycoproteins and lipids are synthesized; Golgi vesicles, which direct membrane constituents to appropriate places in the cell; and peroxisomes, in which fatty acids and amino acids are degraded.

Animal cells, but not plant cells, contain lysosomes, which degrade worn-out cell constituents and foreign materials taken in by the cell. Chloroplasts, where photosynthesis occurs, are found only in certain leaf cells of plants and some single-celled organisms. Both plant cells and some single-celled eukaryotes contain one or more vacuoles, large, fluid-filled organelles in which nutrients and waste compounds are stored and some degradative reactions occur.

The cytosol of eukaryotic cells contains an array of fibrous proteins collectively called the cytoskeleton. Three classes of fibres compose the cytoskeleton: microtubules (20 nm in diameter), built of polymers of the protein tubulin; microfilaments (7 nm in diameter), built of the protein actin; and intermediate filaments (10 nm in diameter), built of one or more rod-shaped protein subunits. The cytoskeleton gives the cell strength and rigidity, thereby helping to maintain cell shape. Cytoskeletal fibres also control movement of structures within the cell; for example, some cytoskeletal fibres connect to organelles or provide tracks along which organelles move. The rigid cell wall, composed of cellulose and other polymers, that surrounds plant cells contributes to their strength and rigidity. Fungi are also surrounded by a cell wall, but its composition differs from that of bacterial or plant cell walls.

CELLULAR DNA

The DNA in the nuclei of eukaryotic cells is distributed among 1 to more than 50 long linear structures called chromosomes. The number and size of the chromosomes are the same in all cells of an organism, but vary among different types of organisms. Each chromosome comprises a single DNA molecule associated with numerous proteins, and the total DNA in the chromosomes of an organism is referred to as its genome. Chromosomes, which stain intensely with basic dyes, are visible in the light microscope only during cell division when the DNA becomes tightly compacted. In all prokaryotic cells, most of or all the genetic information resides in a single circular DNA molecule, about a millimeter in length; this molecule lies, folded back on itself many times, in the central region of the cell. Although the large genomic DNA molecule in prokaryotes is associated with proteins and often is referred to as a chromosome, the arrangement of DNA within a bacterial chromosome differs greatly from that within the chromosomes of eukaryotic cells.

THE LIFE CYCLE OF CELLS

A cell in an adult organism can be viewed as a steady-state system. The DNA is constantly read out into a particular set of mRNAs, which specify a particular set of proteins. As these proteins function, they are also being degraded and replaced by new ones, and the system is so balanced that the cell neither grows, shrinks, nor changes its function. This static view of the cell, however, misses the all-important dynamic aspects of cellular life.

The dynamics of a cell can best be understood by examining the course of its life. A new cell arises when one cell divides or when two cells, like a sperm and an egg cell, fuse. Either event sets off a cell-replication programme that is encoded in the DNA and executed by proteins. This programme usually involves a period of cell growth, during which proteins are made and DNA is replicated, followed by cell division, when a cell divides into two daughter cells. Whether a given cell will grow and divide is a highly regulated decision of the body, assuring that an adult organism replaces worn out cells or makes more cells in response to a new need. Examples of the latter are the growth of muscle in response to exercise or damage, and the proliferation of red blood cells when a person ascends to a higher altitude and needs more capacity to capture oxygen. However, in one major and devastating disease — cancer — cells multiply even though they are not needed by the body. To understand how cells become cancerous, biologists have intensely studied the mechanisms that control the growth and division of cells.

THE CELL CYCLE FOLLOWS A REGULAR TIMING MECHANISM

Most eukaryotic cells live according to an internal clock; that is, they proceed through a sequence of phases, called the cell cycle, during which DNA is duplicated during the synthesis (S) phase and the copies are distributed to opposite ends of the cell during mitotic (M) phase. Progress along the cycle is controlled at key checkpoints, which monitor the status of a cell, for instance, the internal amount of DNA or the presence of extracellular nutrients. When certain conditions are met, the cell proceeds to the next checkpoint. The cycle begins after the cell divides into two daughter cells, each containing an identical copy of the parental cell's genetic material.

The cell cycle of prokaryotes is simple and fast. Replication of the single chromosome begins at a particular DNA sequence, the replication origin, which is anchored to the cell membrane. Once DNA replication is complete, assembly of new membrane and cell wall forms a septum, which eventually divides the cell in two. Because the origins of the two newly formed chromosomes are anchored to different membrane sites, each daughter cell receives one chromosome. In ideal growth conditions, the bacterial cell cycle is repeated every 30 minutes.

Only a few types of eukaryotic cells can grow and divide as quickly as bacteria. Most growing plant and animal cells take 10 – 20 hours to double in number, and some duplicate at a much slower rate. Many cells in adult animals, such as nerve cells and striated muscle cells, do not divide at all. They have temporarily exited from the cell cycle after mitosis and entered a "paused or quiescent" state called G_0. Because eukaryotic cells are larger and more complex than prokaryotic cells, a specialized mechanism coordinates their replication of genomic DNA, distribution of chromosomes, and cell division.

MITOSIS

Mitosis is the mechanism in eukaryotes for partitioning the genome equally at cell division. To accomplish this complex task, plant and animal cells build a specialized machine, called the mitotic apparatus, which captures the chromosomes and then pushes and pulls them to opposite sides of the dividing cell. Remarkably, the mitotic apparatus is a temporary structure that exists only during mitosis to distribute the genetic material. Although the events of mitosis unfold continuously, they are conventionally divided into four substages representing phases of chromosome movement. During the first substage, prophase, the replicated chromosomes, each comprising two identical chromatids, are condensed into compact packets and then released to the cytoplasm when the nuclear membrane breaks down. During metaphase and anaphase, the chromosomes are sorted, and each chromatid of a pair moves to opposite sides of the cell. The end of mitosis is marked by re-formation of a membrane around each set of chromosomes (telophase). Division of the cytoplasm, called cytokinesis, then yields two daughter cells, each with a $2n$ complement of genetic material.

Cell division in plant and animal cells differs mainly at cytokinesis. Animal cells divide in two by pinching of the cytoplasm. However, because a plant cell is surrounded by a rigid cell wall, daughter cells are formed by building a new cell membrane and cell wall between the two daughter nuclei, thereby cutting the cytoplasm into two portions.

CELL DIFFERENTIATION CREATES NEW TYPES OF CELLS

The most complicated example of cellular dynamics occurs when a cell changes, or *differentiates,* to carry out a specialized function. This process often is marked by a change in the microscopic appearance, or *morphology,* of the cell. For example, the different structures of a nerve cell and a muscle cell reflect their respective functions in long-distance communication and contraction, highlighting the biological principle that "form follows function."

Cell differentiation creates the diversity of cell types that arise during the development of an organism from a fertilized egg. This is a process of extensive cell multiplication and differentiation. A mammal that starts as one cell becomes an organism with hundreds of diverse cell types such as muscle, nerve, and skin.

Here we see at its most dramatic the power of DNA to control cellular behaviour: development is a DNA-orchestrated set of cellular changes (easily tens of thousands of them) that occur virtually without fail. The almost perfect resemblance of "identical" twins is a testament to the programme encoded by DNA to reproducibly direct the development of a human being.

Nowhere is the variety of cellular activities and responses better illustrated than in the body's immune system. It is there that many cell types come together in organized tissues specifically designed to allow the body to distinguish its own cells from those of foreign invaders.

Within the immune system, we see both development of specialized cells that can recognize invading cells and formation of tissues from cells that originate in various parts of the body. The immune-system cells not only actively survey their environment with surface receptor proteins like antibodies, but also change their properties when they encounter a foreign substance, allowing the body to rid itself of invaders.

CELLS DIE BY SUICIDE

Unchecked cell growth and multiplication produce a mass of cells, a tumor. *Programmemed cell death* plays the very important role of population control by balancing cell growth and multiplication. In addition, cell death also eliminates unnecessary cells. For example, during embryogenesis, the digits of our fingers and toes are sculpted by the death of cells in the intervening spaces. If these cells remained alive, our hands and feet would become webbed. Thus the timing and location of cell death, as well as cell growth and division, must be precisely controlled.

Cell death follows an internal programme of events called apoptosis, in which all traces of a cell vanish. The first visible sign of apoptosis is condensation of the nucleus and fragmentation of the DNA. The cell soon shrivels and is consumed by macrophages. A cell is directed to commit suicide when an essential factor is removed from the extracellular environment or when an internal signal is activated.

Thus, the default state of the cell is to remain alive. The discovery of genes that suppress the growth of tumors by activating cell death stimulated an exciting new line of cancer research that may lead to more effective treatment strategies

INTEGRATING CELLS INTO TISSUES

The evolution of multicellular organisms permitted specialized cells and tissues to form; a flowering plant has at least 15 cell types, and a vertebrate hundreds. In both plants and animals, cells that are specialized to carry out a particular task are found together in the tissues in which the task is performed: a xylem or meristem; a liver, a muscle, or a nerve ganglion. Different types of cells in a tissue are often arranged in precise patterns of staggering complexity. For instance, the hundreds of different types of neurons in the human brain are interconnected to one another through a network of some 10^{15} synaptic connections! The coordinated functioning of many types of cells within tissues, and of multiple specialized tissues, permits the organism as a whole to move, metabolize, reproduce, and carry out other essential activities.

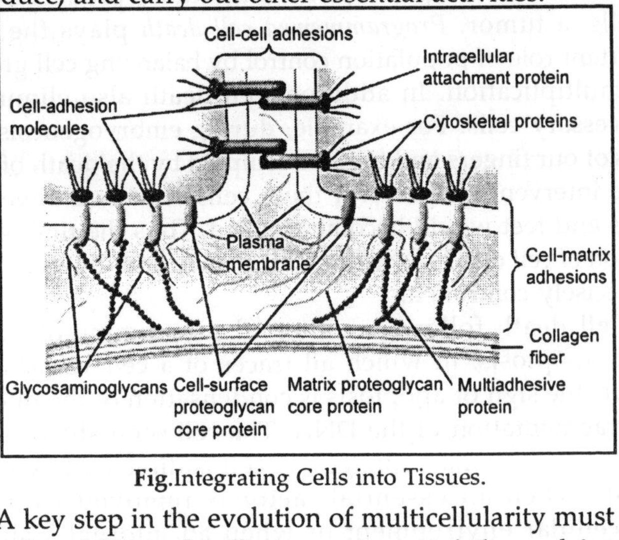

Fig.Integrating Cells into Tissues.

A key step in the evolution of multicellularity must have been the ability of cells to contact tightly and interact specifically with other cells. Various integral membrane proteins, collectively termed cell-adhesion molecules (CAMs), enable many animal cells to adhere tightly and specifically with cells of the same, or similar, type; these interactions allow populations of cells to segregate into distinct tissues. Following aggregation, cells elaborate specialized cell junctions that stabilize these interactions and promote local communication between adjacent cells. Animal cells also secrete a complex network of proteins and carbohydrates, the extracellular matrix (ECM), that creates a special environment in the spaces between cells. The matrix helps bind the cells in tissues together and is a reservoir for many hormones controlling cell growth and differentiation. The matrix also provides a lattice through which cells can move, particularly during the early stages of differentiation. Defects in these connections lead to cancer and developmental malformations.

The extracellular matrix has three major protein components: highly viscous proteoglycans, which cushion cells; insoluble collagen fibres, which provide strength and resilience; and soluble multiadhesive matrix proteins, which bind these components to receptors on the cell surface. Different combinations of these components tailor the strength of the extracellular matrix for different purposes. For example, animals contain many types of extracellular matrices, each specialized for a particular function such as strength (in a tendon), cushioning (in cartilage), or adhesion. In the case of smooth muscle cells that surround an artery, the extracellular matrix must provide strong but flexible connections.

The extracellular matrix is not just an inert framework or cage that supports or surrounds cells. The matrix also communicates directly and indirectly with the intracellular signaling pathways that direct a cell to carry out specific functions. For example, the ability of hepatocytes—the principal cells in the liver—to express liver-specific proteins depends on their association with a matrix of appropriate composition. Specific ECM components can directly activate cytosolic signal-transduction pathways by binding to cell-adhesion protein receptors in the plasma membrane. Alternatively, by binding growth factors and other hormones, the matrix can either sequester these signals from cells or, conversely, present them to cells, thereby indirectly inducing or inhibiting signaling pathways. Morphogenesis—the later stage of embryonic development during which form is achieved by cell movements and rearrangements—also is critically dependent on ECM components, which are constantly being remodeled, degraded, and resynthesized locally. Even in adults—in areas of wounding, for example—degradation and resynthesis of ECM components occurs.

2

Components of Cell

Cellular component refers to the unique, highly organized substances and substances of which cells, and thus living organisms, are composed. Examples include membranes, organelles, proteins, and nucleic acids. Whilst the majority of cellular components are located within the cell itself, some may exist in extracellular areas of an organism.

Cellular components may also be called biological matter or biological material. Most biological matter has the characteristics of soft matter, being governed by relatively small energies. All known life is made of biological matter. To be differentiated from other theoretical or fictional life forms, such life may be called carbon-based, cellular, organic, biological, or even simply living—as some definitions of life exclude alternative biochemistry.

To know the organization of biological systems, one should first know better the main constituent molecules, particularly those of high molecular weight, such as proteins, nucleic acids, polysaccharides, and lipids.

CHEMICAL COMPONENTS

Biochemical analysis of whole tissues was the early approach to the study of chemical composition of the cell like as the liver, brain, skin, or plant meristem. This method had limited cytologic value, because the material analysed was generally a mixture of different cell types and, in addition, contained extra cellular material.

In recent years the development of cell fractionation methods and various micro methods has led to the isolation of different subcellular particles and thus to more important and precise information about the molecular architecture of the cell.

The chemical components of the cell may be classified as *inorganic* (water and mineral ions) and *organic* (proteins carbohydrates, nucleic acids, lipids, and so forth). The protoplasm of a plant or animal cell contains 75 to 85 per cent water, 10 to 20 per cent protein 2 to 3 per cent lipid, 1 per cent carbohydrates, and 1 per cent inorganic material.

SALTS AND IONS

Salts dissociated into anions (*e.g.*, Cl^-) and cations (*e.g.*, Na^+ and K^+) are important in maintaining *osmotic pressure* and the *acid-base equilibrium* of the cell. Retention of ions produces an increase in osmotic pressure and thus the entrance of water. Some of the inorganic ions, such as magnesium, are indispensable as co-factors in enzymatic activities; others, such as inorganic phosphate, form adenosine triphosphate (ATP), the chief supplier of chemical energy for the living processes of the cell, through oxidative phosphorylation.

The concentration of various ions in the intracellular fluid differs from that in the interstitial fluid. For example, the cell has a high concentration of K^+ and Mg^{2+}, while Na^+ and Cl^- are localized mainly in the interstitial fluid. The dominant anion in cells is phosphate; some bicarbonate is also present. *Calcium ions* are exist in the circulating blood and in cells. In bone they combine with phosphate and carbonate ions to form a crystalline arrangement.

Phosphate occurs in the blood and tissue fluids as a free ion, but much of the phosphate of the body is bound in the form of phospholipids, nucleotides, phosphoproteins, and phosphorylated sugars. As primary phosphate (H_2PO_4) and secondary phosphate (HPO_4^{2-}), phosphate contributes to the acid-base equilibrium, thereby buffering the pH of the blood and tissue fluids.

Other ions found in tissues are sulfate, carbonate, bicarbonate, magnesium, and amino acids. Certain *mineral components* are also found as part of larger macromolecules.

For example, *iron,* bound by metal-carbon linkages, is found in hemoglobin, ferritin, the cytochromes, and some enzymes.

WATER

Expect bone and enamel, water is the most abundant component of cells. It serves as a natural solvent for mineral ions and other substances and also as a dispersion medium of the colloid system of protoplasm. For example, from micro-injection experiments it is known that water is readily miscible with protoplasm. Furthermore, water is must require for metabolic activity, since physiologic processes occur exclusively in aqueous media. Water molecules also participate in many enzymatic reactions in the cell and can be formed as a result of metabolic processes.

In cell, water found in two forms: *free* and *bound. Free water* represents 95 per cent of the total cellular water and is the principal part used as a solvent for solutes and as a colloid dispersion medium. *Bound water,* which represents only 4 to 5 per cent of the total cellular water, is loosely held to the proteins by hydrogen bonds and other forces. Because of the asymmetrical distribution of charges a water molecule acts as a *dipole.*

Because of this polarity, water can bind electro statically to both positively and negatively charged groups in the protein. Thus, each amino group in a protein molecule is capable of binding 2.6 molecules of water. Water can also be used to instance eliminate substances from the cell and to absorb heat—by virtue of its high specific heat coefficient—thus preventing drastic temperature changes in the cell.

The water content of an organism is related to the organism's age and metabolic activity. For instance, it is highest in the embryo (90 to 95 per cent) and decreases progressively in the adult and in the aged.

Macromolecules

Structural and other properties of the cell are intimately related to large molecules made of repeating units linked by covalent bonds. Such units are known as monomers and outcome macromolecule is called a *polymer*. Molecules having an increasing number of monomers possess widely different characteristics. For example, among the hydrocarbons, methane and ethane are gases, while butane and octane are liquids; further polymerization (20 or more monomers) produces oils, and finally solids, such as paraffins.

The three basic examples of polymers in living organisms are as follows:

1. *Nucleic acids* result from the repetition of four different units called *nucleotides*. The repetition of the four nucleotides in the DNA molecule is the primary source of genetic information.
2. *Polysaccharides* can be polymers of monosaccharides, forming starch, cellulose, or glycogen, or may also involve the repetition of other molecules, forming more complex polysaccharides.
3. *Proteins* and *polypeptides* consist of the association in various proportions of some 20 different amino acids linked by peptide bonds. The order in which these 20 monomers can be linked gives rise to an astounding number of combinations in various protein molecules. This can determine not only their specificity, but in certain cases, their biological activity.

PROTEINS

The Chains of Amino Acids Linked by Peptide are called Proteins

The building blocks of proteins are the amino acids. Essentially, an amino acid is an organic acid in which the carbon next to the—COOH group (called an alpha carbon) is bound to an amino group ($-NH_2$). In addition, the alpha carbon is bound to a side-chain (R) which is different in the various amino acids.

The amino acids differ from one another only in the side-chain; for example, the R in alanine has one carbon, while in leucine it has four carbons. The properties of the various amino acids depend on the chemical composition, of their side-chains;

for example, lysine and arginine are basic because their sidechains contain an extra amino group, and the acidic amino acids (glutamic and aspartic acids) contain an extra carboxyl group. Because of the simultaneous presence of acidic (carboxyl) and basic (amino) groups, amino acids can have both positive and negative charges and are therefore *amphoteric molecules* or zwitterions.

The ionized form of an amino acid is:

$$^{+}H_3N-\overset{\displaystyle H}{\underset{\displaystyle R}{\overset{|}{\underset{|}{C}}}}-COO^{-}$$

The condensation of amino acids to form a protein molecule occurs in such a manner that the acidic group of one amino acid combines with the basic group of the adjoining one, with the simultaneous loss of one molecule of water.

The linkage—NH—CO—is known as the *peptide linkage* or *peptide bond.* The formed molecule preserves its amphoreric character, since an acidic group is always at one end and a basic group is at the other, in addition to the lateral residues (radicals) that can be either basic or acidic. A combination of two amino acids is *dipeptide;* of three, a *tripeptide.* When a few amino acids are linked together, the structure is an *oligopeptide.* A *polypeptide* contains a large number of amino acids.

The distance between two peptide links is about 0.35 nm. A protein with a molecular weight of 30,000 consisting of 300 amino acid residues, if fully extended, should have a length of 100 nm, a width of 1.0 nm, and a thickness of 0.46 nm.

The term *protein* means that all basic functions in living organisms depend on specific proteins. They constitute the enzymes and the contractile machinery of the cell, and are present in the blood, and other intercellular fluids. Some long-chain proteins, such as *collagen* and *elastin* play an important role in the organization of the extracellular framework of tissues.

Student may refer the biochemistry book for detailed study of the classification of Proteins however, it is important to stress that the properties of proteins vary considerably. For instance, *keratin* and *collagen,* are insoluble and fibrous; the *globular proteins e.g.,* egg albumin and serum proteins, are soluble in water or salt solutions and are spherical rather than threadlike molecules.

The *conjugated proteins* are attached to a nonprotein moiety, the so-called *prosthetic* group. To such a group belong the *nucleoproteins* associated with nucleic acids, the *glycoproteins,* the *lipoproteins* (*e.g.,* blood lipoproteins), and the *chromoproteins,* which have a pigment as the prosthetic group, such as hemoglobin, haemocyanin, and the cytochromes. Hemoglobin and myoglobin (present in muscle) contain the prosthetic group *heme,* an iron-containing organic compound that combines with oxygen.

Primary Structure

The amino acid sequence of the polypeptide chain is known as the *primary structure* of the protein molecule. It is the most important and specific structure and determines the so-called *secondary* and *tertiary* structures. Groups of protein subunits containing secondary and tertiary structures constitute the *quaternary* structure; many proteins are formed by several polypeptide subunits. In the protein molecule, amino acids are arranged like beads on a string, and their sequence is of great biological importance. In the hemoglobin molecule a change in a single amino acid produces profound biological changes.

Secondary Structure

In a protein made by several hundred amino acids, the chain may sometimes be linear, but more frequently it assumes different shapes that constitute the secondary *structure.* Fibrous proteins are often arranged in an orderly manner that can be analysed by x-ray diffraction methods. This technique has facilitated classification of proteins into three structural types or groups.

The β-*keratin type* has an identity period of about 0.72 nm, and the adjacent chains are disposed in a *pleated sheet structure.* In which the side-chains of the amino acid residues stick out perpendicular to the plane of the chain. By hydrogen bond the individual chains are chains are held together which formed peptide grid

The α-*helix structure* found in the *a-keratin type* is produced when the polypeptide chain forms a helical structure, like a spiral winding around an imaginary cylinder, in such a way that hydrogen bonds are established within the molecule and not with an adjacent molecule. For the *collagen group,* a model made of three helical chains has been proposed.

Tertiary Structure

In the d globular *proteins* the polypeptide chain is held together in a definite way to form a compact structure in three dimensions. The arrangement in space of such chains is very complex but may be resolved by X-ray diffraction. In globular proteins the chains are folded in a compact way with the polar groups towards the surface thus leaving little space in the interior for water molecules (the hydrophobic amino acids are in the centre of the molecule). These proteins have regions of a -helix or b configurations, and other parts exist as a *random coil, i.e.,* in a flexible structure that, may change at random. The tertiary structure is given by the way in which these helical or random-coil segments arrange with respect to each other.

The spatial arrangement is predetermined by the sequence of amino acids in the primary structure and by the bonds that can be established among some of the residues. Enzyme activity and antigenicity *i.e.,* series of biological properties are related to the tertiary structure.

The *denaturation* of a protein is brought about by high temperatures or other nonphysiologic conditions, and consists of a disruption of the tertiary structure. This is usually accompanied by loss of biological activity.

In the case of the enzyme ribonuclease, denaturation can be achieved by treatment with b-mercaptoethanol and high concentrations of urea. Mercaptoethanol is a reducing agent, which can disrupt—S—S—bridges (disulfide bonds), reducing them to —SH groups; while urea disrupts weak interactions (see later discussion). The tertiary structure of ribonuclease is held together by four disulfide bonds, which are established between pairs of cysteines (an amino acid that contains an —SH group).

After denaturation, the enzyme can be correctly refolded into its natural conformation *(renaturation) by* removing the urea and mercaptoethanol gradually by dialysis; after this, the enzymatic activity is recovered. There are 105 possible combinations in which eight cysteines can pair to produce four disulfide bridges, but only the biologically active natural conformation is produced after careful renaturation, because it is thermodynamically the most stable structure. This is a clear demonstration that all the information needed to produce the complex folding of a protein molecule is contained in its amino acid sequence.

Quaternary Structure

Protein Subunits: The primary, secondary, or tertiary structures, are concern a single polypeptide chain but quaternary structure contain two or more chains These chains may or may not be identical, but in both cases they are linked by weak bonds. For example, the hemoglobin molecule is composed of four polypeptides or subunits, two designated as a and two as b. Separation and association of the subunits may occur spontaneously. Hemoglobin may be broken into two half molecules (two a and two b) by urea. When urea is removed, they reassemble, forming complete, functional molecules. This binding is highly specific and takes place only between the half molecules. This is called the *principle of selfassembly*. This principle also applies to the building up of more complex cellular structures, such as the cell membrane, microtubules, and so forth.

Weak Interactions-Essential for Protein Structure

In the structure of protein many types of bonds are involved The primary structure is fully determined by *covalent bonds* (peptide bonds).

Disulfide bonds (—S—S—bridges) are also covalent and are established between the —SH groups of two cysteine residues.

Disulfide bonds can be reversibly dissociated by the action of *reducing* agents (such as mercaptoethanol and dithiothreitol-DTT). In addition, several *weak interactions* are very important in the establishment of the secondary and tertiary structures. All these weak bonds are non-covalent; the main types are: *Ionic or electrostatic bonds,* which result from the attractive force between ionized groups of opposite charge.

Hydrogen bonds result from a H^+ (proton) that is shared between two neighbouring electronegative atoms. The H^+ can be shared between nitrogen or oxygen atoms which are close to each other.

Hydrogen bonds' are very important in biology and are the main force that holds the two strands of DNA together. Hydrogen bonds are essential for the specific pairing between nucleic acid bases, which is the basis for the coding of genetic information.

In *hydrophobic interactions* water tends to be excluded by non-polar groups, which associate with each other so that they are not in contact with water. In globular proteins the sidechains of the most hydrophobic amino acids (those with long non-polar side-chains, such as leucine, isoleucine, valine, and phenylalanine) tend to aggregate inside the molecule, and the polar (charged) groups protrude from the surface of the tertiary structure. The hydrophobic residues tend to repel the water molecules that surround the protein, thereby causing the globular structure to be more compact.

Van der Waals interactions occur only when two atoms come Very close together. The closeness of two molecules can induce charge fluctuations which may produce dipoles and mutual attraction at very short range.

The essential difference between a covalent and a non-covalent bond is in the amount of energy needed to break the bond d. Although each individual bond is weak, large numbers of them can produce very stable structures, as in the case of doublestranded DNA. A more detailed discussion will show that covalent bonds are generally broken by the intervention of enzymes, whereas non-covalent bonds are easily dissociated by physicochemical forces.

Electric Charges of Proteins

In addition to the terminal $-NH_3^+$ and$-COO$ charged groups, proteins contain dicarboxylic- and diamino-amino acids which dissociate as follows:

1. The acidic groups lose protons and become negatively charged. For example, in aspartic and glutamic acids, the free carboxyl group dissociates into$-COO + H^+$
2. The basic groups, by gaining protons, become positively charged$-NH_2 + H^+ -NH_3^+$. This is found in amino acids with two basic groups, such as lysine or arginine. AH these so-called *ion-producing groups* contribute to the acid-base reactions of proteins and to the electrical properties of protein molecules.

The actual charge of a protein molecule is the result of the sum of all single charges. Because dissociation of the different acidic and basic groups takes place at different hydrogen ion concentrations of the medium, pH greatly influences the total charge of the molecule. In an acid medium, amino groups capture hydrogen ions and react as bases ($-NH_2 + H^+ \rightarrow -NH_3^+$); in an alkaline medium the reverse takes place and carboxylic groups dissociate ($-$ $COOH \rightarrow COO^- +$ H^-). For every protein there is a definite pH at which the sum of positive and negative charges is zero. This pH is called the *isoelectric point* (pI) At the isoelectric point, proteins placed in an electric field do not migrate to either of the poles, whereas at a lower pH they migrate to the negative pole *(cathode)* and at a higher pH, to the positive pole *(anode)*. This migration is called *electrophoresis,* and it provides a very useful technique for the separation of cellular proteins.

Separation of Cell Proteins

Each protein has a characteristic isoelectric point, and this property can be used in the separation of proteins. In the technique called *isoelectric focusing,* proteins are subjected to electrophoresis on a pH gradient. Each protein moves until it reaches a pH equal to its individual isoelectric point. At that moment, migration in the electric field stops because the net charge of the protein is zero. Recently the techniques of isoelectric focusing and polyacrylamide gel electrophoresis have been combined to produce *two dimensional* separation of proteins. Several hundred cellular proteins can be resolved from one another.

This technique is increasingly used in Cell Biology, and its great resolving power is due to the use of two *independent properties* of proteins. The proteins are first separated by isoelectric focusing (this is the first dimension), which separates proteins according to their *charge* (isoelectric point).

The proteins are subsequently separated by electrophoresis (this is the second dimension) in polyacrylamide gels containing SDS, which separates proteins according to their *size* (molecular weight). This technique results in a series of spots distributed throughout the polyacrylamide gel (if the same property of proteins had been used in both dimensions, the spots would be distributed along a diagonal).

When the detergent sodium dodecyl sulfate (SDS) is used with electrophoresis, the proteins are separated mainly according to their molecular weight. This is because SDS binds to the proteins, giving them large numbers of negative charges due to the sulfate. Thus, most of the protein charges will come from the SDS, minimizing the role of charge differences between individual proteins (differences which would otherwise affect electrophoretic mobility), and all the proteins migrate according to their size.

The larger proteins move more slowly than the smaller ones because they encounter more resistance when traversing the mclecular pores within the polyacrylamide gel used for electrophoresis. SDS electrophoresis is widely used as a method for determining molecular weights of proteins.

CARBOHYDRATES

Carbohydrates which is known as the energy food composed of carbon, hydrogen, and oxygen, are sources of energy for animal and plant cells; in many plants they also form important constituents of cell walls and serve as supporting elements. Animal tissues have fewer carbohydrates; among the most important are glucose, galactose, glycogen, and amino sugars and their polymers.

Monosaccharides, disaccharides, and polysaccharides are the carbohydrater of biological importance. The first two, commonly referred to as *sugars,* are readily soluble in water, can be crystallized, and easily pass through dialyzing membranes. Polysaccharides, on the other hand, neither crystallize nor pass through membranes.

MONOSACCHARIDES

Monosaccharides are simple sugars having the empirical formula $C_n(H_2O)_n$. They are classified in accordance with the number of carbon atoms, *e.g.,* trioses, pentoses, and hexoses. The pentoses *ribose* and *deoxyribose* are found in the molecules of nucleic acids, and the pentose *ribulose* is important in photosynthesis. Glucose, a hexose is the primary source of energy for the cell. Other important hexoses are *galactose* and *fructose (levulose).*

DISACCHARIDES

Disaccharides are sugars formed by the condensation of two monomers of monosaccharides with the loss of one molecule of water. Their empirical formula is therefore $C_{12}H_{22}O_{11}$. The most important of this group are *sucrose* (formed by glucose and fructose) and *maltose* (formed by two glucose monomers) in plants, and *lactose* (formed by galactose and glucose) in animals.

POLYSACCHARIDES

Polysaccharides result from the condensation of many molecules of monosaccharides, with a corresponding loss of water molecules. Their empirical formula is $(C_6H_{10}O_3)_n$. Upon hydrolysis they yield molecules of simple sugars. The most important polysaccharides in living organisms are *starch* and *glycogen,* which are reserve substances in cells of plants and animals, respectively, and *cellulose,* the most important structural element of the plant cell.

Starch is a combination of two long polymer molecules: *amylose* and amylopectin. Glycogen may be considered to be the starch of animal cells. The structure of branched polymer, which is composed of many molecules of glucose. It is found in numerous tissues and organs, but the greatest proportion is contained in liver cells and muscle fibres.

Complex Polysaccharides

In addition to the polysaccharides made of hexose monomers mentioned in the preceding section, there are many more complex long molecules that contain amino nitrogen (*e.g.*, glucosamine) that can, in addition, be acetylated (*e.g.*, acetylglucosamine) or substituted with sulphuric or phosphoric acid.

All these polymers are important in molecular organization, particularly as intercellular substances. These polysaccharides may exist either freely or combined with proteins, or lipids.

The most important are: *Neutral polysaccharides,* which contain only, acetylglucosamine. The main example is *chitin,* a supporting substance found in the exoskeletons of insects and crustaceans.

Acidic mucopolysaccharides, which contain sulphuric or other acids in the molecule. These molecules are strongly basophilic. To this group belong *heparin,* an anticoagulant substance; *chondroitin sulfate,* present in the cartilage, skin, cornea, and umbilical cord; and *hyaluronic acid,* in skin and other animal tissues. The latter is hydrolyzed by *hyaluronidase.*

Glycoproteins

Glycoproteins are complexes composed of a protein and a prosthetic group of carbohydrates. Several monosaccharides, such as galactose, mannose, and fucose, as well as N-aceryl-Dglucosamine and sialic acid, are found.

Glycoproteins can be divided into two major categories: cellular ana secretory. Cellular glycoproteins are present mainly in the cell membrane, with the carbohydrate moiety protruding to the outside of the cell, and they have important functions in membrane interaction and recognition. Most secreted proteins are glycoproteins and are produced by various cells: serum glycopro-, teins (*i.e.*, seroalbumins), secreted by the liver; thyroglobulin, produced in the thyroid gland, immunoglobulins, secreted by plasma cells; ovalbumin, secreted by the hen oviduct; and ribonuclease and deoxyribonuclease, produced by the pancreas. In most glycoproteins the protein is linked to the carbohydrate moiety by a bond between asparagine (Asn) and N-acetyl-Dglucosamine (GlcNAc). Evidence obtained by Leloir's group at Buenos Aires indicates that the oligosaccharide core of glycoproteins is first assembled on a lipid carrier and is then transferred to the protein molecule. The function of the lipid carrier (a derivative of *dolichol)* is to enable the hydrophilic oligosaccharide to traverse the endoplasmic reticulum membrane.

Several exocrine glands, such as the salivary glands and the mucous glands of the digestive tract, secrete mucoproteins in which the linkage is between N-acetylglucosamine and serine or threonine in the protein. Sialic acid and L-fucose are the terminal sugars in these compounds. The carbohydrate moiety of glycoproteins is important in determining certain physicochemical properties (the viscosity of mucins may be important in protecting and lubricating the mucous membrane of the digestive tract, for example); in molecule-membrance interactions (for the determination of blood groups and other immunological properties at the surface of erythrocytes); and in membrane-membrane interactions.

LIPIDS

The compounds in this large group relative insolubility in water and solubility in organic solvents. The cause of this general property of lipids and related compounds is the-predominance of long aliphatic hydrocarbon chains or benzene rings. These structures are non-polar and hydrophobic. In many lipids these chains may be attached at one end to a polar group, rendering it capable of binding water by hydrogen bonds. Lipids can be classified in the following way.

Triglycerides

Simple lipids are alcohol esters of fatty acids these are:

- *Neutral fats* (glycerides), it is called triglycerides they are triesters of fatty acids and glycerol. Neutral fats accumulate in adipose tissue.
- *Waxes,* which have a higher melting point than natural fats; they are esters of fatty acids with alcohols other than glycerol, such as in beeswax.

Fatty acids have long hydrocarbon chains Fatty acids always have an even number of carbons (because they are synthesized from two-carbon acetyl units). Some times the hydrocarbon chain has double bonds (—C=C—), and in this case the fatty acid is said to be non-saturated. The carboxyl groups of fatty acids react with the alcohol groups of glycerol.

The outcoming triglycerides are used by organisms to store spare energy conveniently: The oxidation state of the long hydrocarbon chains is very low, and they therefore liberate large amounts of energy (about twice as many calories per gram as carbohydrates and proteins) when oxidized to form CO_2 and water.

Compound Lipids—Phospholipids and Biological Membranes

Upon hydrolysis these lipids yield other compounds in addition to alcohol and acids. They serve mainly as structural components of the cell, particularly in cell membranes. The following are classified as compound lipids:

Phosphatides (phospholipids) are diesters of phosphoric acid that can be esterified with either glycerol or sphingosine. Lecithins, cephalins, inositides and plasmalogens are included in this group A phospholipid has only two fatty acids attached to the glycerol molecule. The third hydroxyl group of glycerol is esterified to *phosphoric acid* instead of to a fatty acid. This phosphate is also bound to a second alcohol molecule, which can be *choline, ethanolamine, inositol,* or *serine* depending on the type of phospholipid: Phospholipids have two long *hydrophobic* fatty-acid "tails" and a *hydrophilic* (polar) phosphate-containing "head." Phospholipids are *amphipathic* molecules (molecules with a hydrophilic and a hydrophobic region), and they are the basic constituents of biological membranes. Biological membranes are bilayers of phospholipids with the hydrophilic heads (phosphate-containing regions) positioned at the water interface and the long hydrophobic tails avoiding water and arranged in the interior of this membrane structure.

Glycolipids and *sphingolipids* are characterized by the fact that glycerol is replaced by the amino alcohol *sphingosine.* To these groups belong the *sphingomyelins,* mainly in the myelin sheath of nerves; the *cerebrosides,* which are characterized by the presence of galactose or glucose in the molecule; the *sulfatides,* which contain sulfuric acid esterified to galactose; and the *gangliosides*

The gangliosides need special mention because of their presence in cell membranes, their possible role as receptors of virus particles, and their influence on ion transport across membranes.

A ganglioside is a complex molecule containing sphingosine, fatty acids, carbohydrates (lactose + galactosamine), and neuraminic acid.

NUCLEIC ACIDS

Macromolecules of the utmost biological importance are called Nucleic acids. All living organisms contain nucleic acids in the form of deoxyribonucleic acid (DNA) and ribonucleic acid (RNA). Some viruses may contain only RNA, while others have only DNA.

DNA is the major store of genetic information. This information is transmitted by transcription into RNA molecules which are utilized in the synthesis of proteins.

Most importantly the central dogma of modern biology is:

DNA → RNA → PROTEIN

Only those features of their chemical structure relevant to the understanding of their function will be considered here. In higher cells DNA. is localized mainly in the nucleus as part of the chromosomes. A small amount of DNA is present in the cytoplasm and contained mitochondria and chloroplasts. RNA in found both in the nucleus, where it is synthesized, and in the cytoplasm, where the synthesis of proteins occurs.

Table. Dna and Rna: Structure, Reactions, and Role in the Cell

	Deoxyribonucleic Acid	Ribonucleic Acid
Localization	Primarily in nucleus; also in mitochondria and chloroplasts	In cytoplasm, nucleolus, and chromosomes
Pyrimidine bases	Cytosine Thymine	Cytosine Uracil
Purine bases	Adenine Guanine	Adenine Guanine
Pentose	Deoxyribose	Ribose
Cytochemical Reaction	Feulgen	Basophilic dyes with ribonuclease treatment
Hydrolyzing enzyme	Deoxyribonuclease (DNase)	Ribonuclease (RNase)
Role in cell	Genetic information	Synthesis of proteins

Nucleic Adds

Nucleic acids are made up of sugar moiety (pentose), nitrogenous bases (purines and pyrimidines), and phosphoric acid. After a mild hydrolysis the nucleic acids are decomposed into nucleotides.

Nucleotides are the monomeric units of the nucleic acid macromolecule. Nucleotides result from the covalent bonding of a phosphate and a heterocyclic base to the pentose. Within the nucleotide, the combination of a base with the pentose constitutes a *nucleoside*. For instance *adenine* is *a* purine base; *adenosine* (adenine + ribose) is the corresponding nucleoside and *adenosine monophosphate* (AMP), *adenosine diphosphate* (ADP), and *adenosine triphosphate* (ATP), are nucleotides. In addition to constituting the building blocks of nucleic acids, nucleotides are important because they are used to store and transfer chemical energy.

The two terminal phosphate bonds of ATP contain high energy; when these bonds are cleaved, the energy released can be used to drive a variety of cell functions. Nucleic acids are linear polymers in which the nucleotides are linked together by means of phosphate-diester bridge's with the means moiety. These bonds link the 3' carbon in one nucleotide to the 5' carbon in the pentose of the adjacent nucleotide. The backbone of nucleic acids consists, therefore, of alternating phosphates and pentoses. The nitrogenous bases are attached to the sugars of this backbone.

Phosphoric acid uses two of its three acid groups in the *3′ 5′* diester links. The remaining negative group confers to the polynucleotide its acid properties and enables the molecule to form ionic bonds with basic proteins. In eukaryotic cells, DNA is associated with *histories* (*i.e.,* basic proteins rich, in arginine of lysine), forming a *nucleoprotein complex* called *chromatin.* This anionic group also cause nucleic acids to be highly basophilic; *i.e.,* they stain readily with basic dyes.

Ribose in RNA and deoxyribose in DNA are two types of Pentoses. The only difference between these two sugars is that the oxygen in the, 2' carbon is lacking in deoxyribose. A cytochemical reaction that is specific for the deoxyribose moiety *(Feulgen reaction)* can be used to visualize DNA under the microscope.

In DNA the pyrimidines are *thymine* (T) and *cytosine* (C); the purines are *adenine* (A) and *guanine* (G). RNA contains uracil (U) Instead of thymine. Therefore *between RNA and DNA there are two main differences:* the first one in the pentose moiety (ribose and deoxyribose, respectively) and the second in a pyrimidine base (uracil instead of thymine). This explains why in Cell Biology radioactive thymidine (*i.e.,* the nucleoside) is used to label specifically DNA and radioactive uridine for RNA. At 260 nm. Heterocyclic bases the ultraviolet light. A cell photographed at this wavelength shows the nodulous, the chromatin, and the RNA-containing regions of the cytoplasm absorbing intensely. In fact, the simplest way of determining the concentration of a solution of nucleic acids is to measure its absorbance at 260 nm.

All the genetic information of a living organism is stored in the linear sequence of the four bases. Therefore, a four letter alphabet (A, T, C, G) must code for the primary structure of all proteins (*i.e.*, composed of 20 amino acids). All the excitement in molecular biology, leading to the unravelling of the genetic code began when the structure of DNA was understood.

Regularities in DNA Base Composition-A T and G = C

DNA is present in living organisms as linear molecules of extremely high molecular weight. In higher organisms the amount of DNA may be several thousand times larger; for example, the DNA contained in a single human diploid cell, if fully extended, would have a total length of 1.7 meters.

The study about the base composition of DNA was carried by chargeoff. He found that the base composition varies from one species to another, and yet striking regularities were also found. In all cases the amount of adenine was found to be equal to the amount of thymine (*i.e.*, A = T). The number of cytosine and guanine bases were also found to be equal (*i.e.*, G = C). As a consequence, the total quantity of purines equals the total quantify of pyrimidines. (*i.e.*, A + G = C + T). On the other hand, there is considerable variation between species regarding the AT/GC ratio. In higher plants and animals in general AT is in excess of GC, whereas in viruses, bacteria, and lower plants the contrary is more common. For example, in man the AT/GC ratio is 1.52; in *E. coli*, it is 0.93.

DNA

On the DNA model the contribution of Watson and crick is very historical in bioscience. In 1953, based on the x-ray diffraction data of Wilkins and Franklin, Watson and Crick proposed a model for the DNA structure that provided an explanation for the above-mentioned regularities in base composition and for the biological properties of DNA—particularly its duplication in the cell. It is composed of two right-handed helical polynucleotide chains that form a *double helix* around the same central axis. The two strands are *antiparallel*; *i.e.*, their 3',5' phosphodiester links are in opposite directions. Furthermore, the bases are stacked inside the helix in a plane perpendicular to the helical axis.

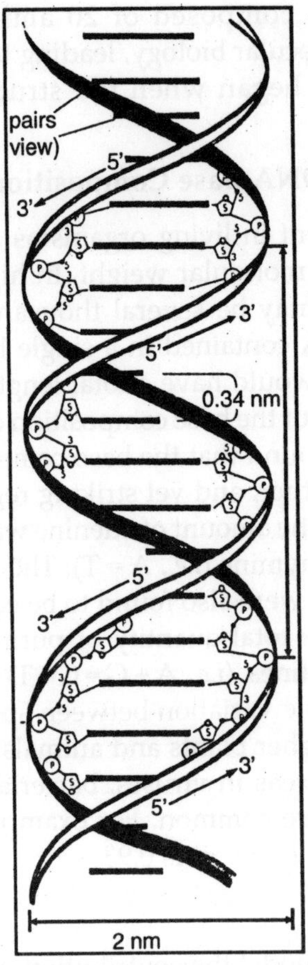

Fig. The DNA double helix.

The two strands are held together by *hydrogen bonds* established between the pairs of bases. Since There is a fixed distance between the two sugar moieties in the opposite strands only certain base pairs can fit into the structure. Only two pairs that are possible are AT and CG. It is important to note that two hydrogen bonds are formed between A and T, and three are formed between C and G, and that therefore a CG pair is more stable than an AT pair. In addition to hydrogen bonds, hydrophobic interactions established between the stacked bases are important in maintaining the double helical structure.

In the Watson and Crick model the distance between the stacked bases is 0.34 nm which corresponds to the primary period demonstrated by x-ray diffraction. Furthermore a turn of the double helix is completed in 3.4 nm, a length which corresponds to 10 nucleotide residues. It is proved from observation of a space-filling model of DNA that the double helix has a mean diameter of 2.0 nm; furthermore, two grooves (a major or deep groove, and a minor or more shallow one) are observed.

The *axial sequence* of bases along one polynucleotide chain may vary considerably, but on the other chain the sequence must be *complementary*. Because of this property, given an order of bases on one chain, the other chain is exactly complementary.

During DNA duplication, the two chains dissociate, and each one serves as a template for the synthesis of two complementary chains. In this way two DNA molecules are produced each having exactly the same molecular constitution.

Denaturation and Annealing of DNA

By weak interactions the structure of double helix of DNA is preserved it is possible to separate the two strands by heating and other treatments (such as alkaline pH). This separation process, known as *melting* or *denaturation of DNA,* can be followed in the spectrophotometer by measuring the absorbance at 260 nm. Since the double helix of DNA absorbs less ultraviolet light than the sum of the, two isolated strands, by raising the temperature it may be observed that at a certain point there is a sharp increase in the quantity of ultraviolet light absorbed sometimes called *hyperchromic shift*).

The increase in temperature at this point corresponds to the "melting" (or separation) of two DNA strands and is variable for DNAs of different base composition. Since the temperature required to break the GC pair (having three hydrogen bonds) is higher than the one needed to break the AT pair (having two hydrogen bonds), the melting point of a given DNA sample will depend on the AT/GC ratio. If after denaturation the DNA is cooled slowly the complementary strands will base pair in register, and the *native* (double helical) conformation will be restored. This process is called *renaturation* or *annealing*.

Renaturation of DNA is a very useful tool in Molecular Biology. Shows how DNA, renaturation can be used to estimate the size (number of nucleotides) of the genome of a given organism. When DNA is renatured under standardized conditions, a large genome (*e.g.*, calf) takes more time to reanneal than a small genome (*E. coli* or bacteriophage T_4). This is because the individual sequences take longer to find the correct partners (the larger the genome, the more chances there are of incorrect molecular collisions):

Renaturation studies have led to the discovery of *repeated sequences* in eukaryotic DNA. When certain copies are repeated many times, the rate of renaturation will be much faster than for sequences present as single copies. Some sequences (called *satellite* DNAs) can be repeated millions of times in the genome (10×10^6 times for mouse satellite).

Single-stranded DNA will also anneal to complementary RNA. Molecular *hybridization* is used extensively in Molecular Biology. Purified RNAs (such as, for example, globin mRNA) can be labeled. These labeled probes can then be used to study the genes from which they arose.

Denaturation done under carefully controlled conditions may be used for physical mapping of DNA in a technique known as partial *denaturation mapping*. This technique is based on the fact that the regions rich in AT separate more easily. Under the electron microscope those regions are detected as single-stranded loops, and the distance between the loops and the end of the DNA molecule can be measured.

Circular DNA-Supercoiled Conformation

Many viruses have covalently closed circular DNA molecules that when visualized under the electron microscope show the twisted structure. This DNA is *supercoiled,* which indicates that the structure is under tension. Supercoiling arises from a relative lack of, turns in the DNA double helix. Let us consider a covalently closed circular DNA molecule with 5000 base pairs. In the *relaxed* conformation, the DNA helix will have 500 helical turns (because. the double helix gives one turn every 10 nucleotides). However, in a supercoiled molecule, the helix will have fewer turns (if it has 499, it will have one superhelical coil; if 490, it will have 10 coils). The molecule is under tension and folds upon itself in the supercoiled from. Because the molecule must be covalently dosed, supercoils are always present in integral numbers; *i.e.*, a molecule can lack one, two, or three turns, but not fractional amounts, such as 1.1, 1.2, or 1.3 (halftwists are not possible because the two DNA chains have opposite 3' to 5' polarities).

Fully supercoiled DNA has about one super helical turn every 200 base pairs. It is thought that supercoiling reflects the fact that in eukaryotic cells DNA is normally not free but bound to histone proteins. *In vivo* the DNA is not under tension and adopts the conformation most favourable for chromatin structure. When the histones are removed during purification of the DNA, the tension becomes apparent, and the molecules supercoil.

RNA Structure

The basic structure of RNA is similar to that of DNA, but in RNA in the place of deoxyribose and thymine, ribose and uracil is present. The base composition of RNA does not follow Chargaff's rules, since RNA molecules are made of only one chain. RNA is synthesized within the nucleus by using only one strand of DNA as template (*i.e.*, RNA is a faithful copy of the information contained in the DNA). There are three major classes of ribonucleic acid: ribosomal RNA (rRNA), transfer RNA (tRNA), and messenger RNA (mRNA). All of them are involved in protein synthesis.

Although each RNA molecule has only a single polynucleotide chain, RNA is not a simple, smooth, linear structure. RNA molecules have extensive regions of complementary in which hydrogen bonds between AU and GC pairs are formed between different regions of the same molecule. In the base-paired regions the RNA molecule adopts a helical structure which is comparable to that of DNA. The compact structure of RNA molecules folded upon themselves has important biological consequences. For example, a sequence that indicates the starting site for protein synthesis might be in a part of the molecule inaccessible to the ribosomes, and will only be expressed if some additional factor unfolded the molecule (this does in fact occur in RNA bacteriophages such as MS2.

The Simplest Infectious Agent

The small organism contains nucleic acid is called viruses. It carries the genetic information, packed inside a cover of protein molecules (called a capsid). Viruses do not have their own machinery ' for protein synthesis but rather use the infected cell as a source of ribosomes, energy, and other requirements for successful multiplication.

Viruses are simple enough, but *virioids* are even simpler. Virioids are infectious agents that attack plant cell and that are made of an RNA molecule only, that is *not* covered, with protein. The compact base-paired structure probably protects the molecule from ribonucleases, which are very active and ubiquitous enzymes but which in general prefer single-stranded RNA.

3

Functions of Cell Membrane

The cell membrane is a biological membrane that separates the interior of all cells from the outside environment. The cell membrane is selectively-permeable to ions and organic molecules and controls the movement of substances in and out of cells. It consists of the phospholipid bilayer with embedded proteins. Cell membranes are involved in a variety of cellular processes such as cell adhesion, ion conductivity and cell signaling and serve as the attachment surface for the extracellular glycocalyx and cell wall and intracellular cytoskeleton. The cell is concerned with a different internal milieu from that of its external environment. For example, the ionic content of animal cells is quite dissimilar from that of the circulating blood.

This difference is maintained throughout the life of the cell by the thin surface membrane, the *cell* or *plasma membrane,* which controls the entrance and exit of molecules and ions. The function of the plasma membrane of regulating this exchange between the cell and the medium is called *permeability.* This membrane can't seen by light microscope, but in some cells it is covered by thicker protective layers that are within the limits of microscopic resolution. For example, most plant cells have a thick cellulose wall that covers and protects the true plasma membrane.

Some animal cells are surrounded by cement-like substances that constitute visible cell walls. Such layers, also called *cell coats,* generally play no role in permeability but do have other important functions. The *"fluid mosaic" model,* dominates present views of membrane organization. With this structure as a base we will consider various aspects of *cell permeability*. Then, in relation to certain *differentiations of the cell membrane* we shall examine the important problem of *intercellular communications*. Finally, together with the study of the *cell coat*, we will analyse some of the mechanisms by which cells *recognize* their partners in a tissue. It will be mentioned that changes in intercellular communication and *molecular recognition* are of paramount importance in the *cancerous transformation of cells*. The cell membrane is becoming one of the fields most fundamental to the cell biology of both the normal and the neoplastic cell.

CHEMICAL COMPOSITION

Regarding chemical composition of the plasma membrane of the human red cell is more than about that of any other cell. Protein represents approximately 52 per cent of its mass, *lipids* 40 per cent, and *carbohydrates* 8 per cent. Oligosaccharides are bound to lipids for example *glycolipids* and, mainly, to proteins for example *glycoproteins)*.

There is a wide variation in the lipids-protein ratio between different cell membranes. Myelin is an exception, in the sense that the lipid predominates; in the other cell membranes there is higher protein lipid ratio. It may be observed that in myelin the area occupied by the protein is insufficient to cover that of the lipids, whereas in a red cell ghost the opposite situation is found.

The main lipid components of the plasma membrane are phospholipids, cholesterol, and galactolipids; their proportion varies in different cell membranes. The major proportion of membrane phospholipids is represented by phosphatidylcholine, phosphatidylethanolamine and sphingomyelin, all of which have no net charge at neutral pH (*i.e., neutral phospholipids)* and tend to pack tightly in the bilayer. This property is also shared by cholesterol.) Five to 20 per cent of the phospholipids are acidic, including: phosphatidylinositol, phosphatidylserine, cardiolipin, phosphatidylglycerol, and sulfolipids. *Acidic phospholipids* are negatively charged and in the membrane are associated principally with proteins by way of lipid-protein interactions. One of the main characteristics of the molecular organization of the plasma membrane is the asymmetry of all of its chemical components. This refers to their non-uniform distribution between the two surfaces, *i.e.*, the inner or *protoplasmic surface* (P_s), in contact with the ground cytoplasm, and the outer or *external surface* (E_s) in contact with the surrounding fluid medium.

Regarding non-permeant reagents and various *phospholipases (i.e.,* enzymes that hydrolyze different parts of the phospholipid molecule), it has been demonstrated that the distribution of the; *phospholipids is* highly asymmetrical. While the *outer layer* consists mainly of lecithin and sphingomyelin, the *inner layer* is composed mainly of phosphatidylethanolamine and phosphatidylserine. Furthermore, the *glycolipids* are mainly in the outer half of the bilayer. It is assumed that this asymmetry is rather stable and that there is no exchange of lipids across the bilayer.

Carbohydrates

The distribution of the oligosaccharides is also highly asymmetrical. In both *glycolipids* and *glycoproteins* they are confined exclusively to the external membrane surface. In erythrocyte membranes, hexose, hexosamine, fucose, and sialic acid are bound mainly to proteins. In fact all The proteins present at the outer surface are glycosylated.

Because of the presence of sialic acid residues, as well as carboxyl and phosphate groups, the outer surface of the membrane is negatively charged; consequently, positively charged proteins may be bound by electrostatic interactions to the plasma membrane. Only a small amount of sialic acid exists in the form of *gangliosides* (*i.e.*, glycolipids) in the plasma membrane of liver. However, gangliosides are important constituents of the neuronal surface and are probably involved in ion transfers.

Membrane Proteins

Proteins represent the main component of most biological membranes. They play an important role, not only in the mechanical structure of the membrane, but also as carriers or channels, serving for transport) they may also be involved in regulatory or ligand-recognition properties. In addition, numerous enzymes, antigens, and various kinds of receptor molecules, are present in plasma membranes.

Membrane proteins have been classified as *integral (intrinsic) or peripheral (extrinsic) according* to the degree of their association with the membrane and the methods by which they can be solubilized.

Peripheral proteins are separated by mild treatment, are soluble in aqueous solutions, and are usually free of lipids. Examples include: the above-mentioned *spectrin*, which may be removed from red cell ghosts by chelating agents; cytochrome *c*, found in mitochondria; and acetylcholinesterase, in electroplax membranes, which are easily removed in high salt solutions.

Integral proteins represent more than 70 per cent of the two protein types and require drastic procedures for isolation. Usually they are insoluble in water solutions and need the presence of detergents to be maintained in a non-aggregated form. The study of integral proteins from different membranes has shown that they are rather heterogeneous in relation to molecular weight. These proteins may be attached to oligosaccharides, thus forming glycoproteins.

Polypeptides of the Red Cell Membrane

The usual method used to separate the membrane polypeptides is to dissolve the erythrocyte ghosts in sodium dodecyl sulfate (SDS) (an ionic detergent) and then to use polyacrylamide gel electrophoresis (PAGE). The gel can be stained for proteins (with Coomassie Blue) or for carbohydrates. Densitometric scan of the PAGE which, by the position of the bands, permits estimation of the molecular weight of the polypeptides and, by the surface of the profile, enables determination of its relative mass. For example, *polypeptides* 1 and 2 corresponding to *spectrin* (*i.e.*, myosin), represent about 30 per cent of the total protein. *Polypeptide* 3 is the major intrinsic protein and represents about 25 per cent.

Protiens molecular organization is highly asymmetrical, and this can be demonstrated by the use of reagents that are unable to cross the membrane. The reagent is first applied to the *intact erythrocyte* and then to the *white ghost* the difference between the two preparations may give information about the position of a particular protein with respect to the outer or inner surface. Some specific labels can be detected on either side of the membrane by using cytochemical and electron microscopic techniques. For example, using antibodies against spectrin or actin (polypeptide 5) it is possible to show that each is in the cytoplasmic surface of the red cell membrane. On the other hand, acetylcholinesterase can be shown to be in the outer surface by its inactivation with proteolytic enzymes.

It can be said with certainty that *every protein constituent is asymmetrically distributed in the red cell membrane.* All readily soluble polypeptides are localized at the cytoplasmic surface, and those that are at the outer surface are tightly bound to the lipid structure of the membrane.

Major Polypeptides of the Red Cell Membrane

On the cytoplasmic side of the red cell membrane there are *polypeptides 1 and 2* (the myosinlike *spectrin*) and *polypeptide 5 (actin).* These proteins are associated into supramolecular structures forming *microfilaments* which can be identified by electron microscopy, especially by scanning electron microscopy. Under the red cell membrane these structures form a filamentous network that gives stability to the membrane by providing a kind of skeletal support to the fluid lipid bilayer, with its intrinsic proteins. This system of microfilaments may also control the characteristic shape of the erythrocyte. Thus, a possible alteration of the microfilaments has been implicated in a hereditary abnormality of the erythrocytes in which they are spheroidal instead of biconcave discs *(hereditary spherocytosis).*

The molecular weight of the major intrinsic protein is 90,000 daltons, spans the thickness of the membrane and has a small amount of carbohydrate on the pole at the outer surface. This polypeptide is present in the membrane as a dimer held together by S-S bonds. There are between 500,000 and 600,000 such dimers per cell, enough to account for the 8 nm particles observed in freeze-fractured membranes. This dimeric protein appears to be involved in the facilitated *diffusion of anions* (*i.e.,* chloride, bicarbonate) across the membrane.

As per the glycophorin has a molecular weight of 55,000, of which 60 per cent is carbohydrate. Near the COOH end of the molecule there is a region that is very hydrophobic and which interacts with the lipids of the membrane. The COOH end is probably exposed to the interior of the red cell. The NH_2 end is more hydrophilic, is exposed to the external environment, and has the attached oligosaccharides that are at the outer surface of the membrane.

At this surface glycoproteins contain protein-bound antigens of the ABO blood groups and other antigens, such as: the MN groups reacting with rabbit antisera, the influenza virus, phytohemagglutinin, and wheat germ agglutinin. It has been calculated that there are some 700,000 copies of glycophorin per human red cell. This protein accounts for 80 per cent of the carbohydrate and 90 per cent of the negatively charged sialic acid present in the cell surface.

Proteolipids among the most hydrophobic integral proteins of the membrane, are characterized by their strong association with lipids and the fact that they are, soluble in organic solvents. First isolated by Folch and Lees from myelin, proteolipids are found in practically all cell membranes, and in many of them they represent receptor proteins for synaptic transmitters or form channels across those membranes.

Asymmetrical Distribution of Enzymes

More than 30 enzymes have been detected in isolated plasma membranes. Those most constantly found are 5'-nucleotidase, Mg^{2+} ATPase, Na^+-K^+ activated-Mg^{2+} ATPase, alkaline phosphatase adenyl, cyclase, acid phosphomono-esterase, and RNAse. Some enzymes have a preferential localization; for example, alkaline phosphatase and ATPase are more abundant at the bile capillaries, while disaccharidases are present in microvilli of the intestine. A specific localization with a mosaic arrangement has been postulated for some of these enzymes. Disaccharidase forms 5 to 6 nm globular units coating the membrane of the microvilli. The plasma membrane lacks the respiratory chain and glycolytic activity.

Of all the membrane-associated enzymes, Na^+-K^+ activated-Mg^{2+} ATPase is one of the most important because of its role in ion transfer across the plasma membrane. This enzyme is dependent on the presence of lipids and is inactivated when all lipids are extracted. Enzymes also show an *asymmetrical distribution*. Such as in the outer surface of erythrocytes there are acetylcholinesterase, nicotinamide-adenine dinucleotidase, and the ouabain binding site of the Na^+ K^+ ATPase. In the inner surface there is NADH-diaphorase, G3PD, adenylate cyclase, protein kinase, and ATPase.

MOLECULAR MODELS OF THE CELL MEMBRANE

On the molecular structure of the membrane before the isolation of plasma membranes, theories were generally based on indirect information. Since substances soluble in lipid solvents penetrate the plasma membrane easily, Overton postulated in. 1902 that the plasma membrane is composed of a thin layer of lipid. In 1926 Gorter and Grendell found that the lipid content of hemolyzed erythrocytes was sufficient to form a double layer of lipid molecules over the entire cell surface. This theory was also supported by electrical measurements that indicated a high impedance at the plasma membrane. The high impedance is because of the faex fact that it is difficult for ions to penetrate a lipid layer.

Tension at a water-oil interface is about 10 to 15 dynes per centimeter, whereas surface tension of cells is almost nil. It has been postulated that the low tension is due to the presence of protein layers on the lipid components. In fact, when a very small amount of protein is added to a model lipid-water system, the surface tension is lowered comparably.

In 1935 Danielli and Davson explained about all these properties. They proposed that the plasma membrane contained a *lipid bilayer,* with protein adhering to both lipid-aqueous interfaces. Other models containing globular lipid micelles, globular proteins, or a combination of both have been proposed a globular-bilayer transition has also been postulated. These globular models do not account satisfactorily for the high electrical impedance.

On the fine structure of the plasma membrane, electron microscopy has thrown some light and has revealed the numerous structural differentiations that this membrane and the underlying cell cytoplasm have in different cell types. To resolve the structure of the plasma membrane, extremely thin sections (~ 20 nm) must be used, otherwise one would observe different orientations of the plasma membrane with respect to the plane of the section.

The membrane appears most thin when it is exactly perpendicular to the plane of the section. Membranes of 6 to 10 mm have been observed at the surface of all cells. The membranes of two cells that are in close contact appear as dense lines separated by a space of 11 to 15 nm, which is strikingly uniform and contains a material of low electron density. This intercellular component can be considered as a kind of cementing substance.

The Unit Membrane–Re evaluation of the EM Image

There are three layer in cell membrane outer dense layers of 2.0 nm and a middle clear one of about 3.5 nm led Robertson to postulate the so-called *unit membrane* model. The electron microscopic image was interpreted along the lines of the Danielli-Davson model, so that the clear layer corresponded to the hydrocarbon chains of the lipids, and the dense layers, to the proteins. This unit membrane model was also extended to all kinds of intracellular membranes. That this concept is an oversimplication which does not account for the many proteins traversing the membrane can be readily seen by observing some high resolution micrographs in which fine bridges cross the unit membrane.

On the basis of some experiments the unit membrane has been re-evaluated recently methods designed to avoid the removal of proteins from membranes during the preparation for conventional electron microscopy or freeze-fracturing. With both methods a granular appearance of the membrane is obtained, which is more in keeping with the fact that a high proportion of the proteins traverse the membrane. According to these findings the unit membrane image is, to a great extent, artifactual.

The Fluid Mosaic Model

The main source of the knowledge of the molecular organization of biological membrane comes from an integration of the data on chemical analysis and from the application of several biophysical techniques.

The important concepts are:

- That the lipid and integral proteins are disposed in a kind of mosaic arrangement;
- That biological membranes are quasi-fluid structures in which both the lipids and the integral proteins are able to perform translational movements within the overall bilayer. The concept of fluidity implies that the main components of the membrane are held in place only by means of non-covalent interactions.

For better study of the molecular organization of the membrane, it is necessary to remember that, not only the lipids, but also many of the intrinsic proteins and glycoproteins of the membrane, are amphipathic molecules. The term *amphipathy,* coined by Hartley in 1936, refers to the presence, within the same molecule, of hydrophilic and hydrophobic groups these amphipathic molecules constitute liquid crystalline aggregates in which the polar groups are directed towards the water phase and the non-polar groups are situated inside the bilayer.

The integral proteins of the membrane are intercalated to a greater or lesser extent into a rather continuous lipid bilayer. This arrangement is based on the fact that these integral proteins are also amphipathic, with polar regions protruding from the surface and non-polar regions embedded in the hydrophobic interior of the membrane.

It is well recognized that a protein of appropriate size or a cluster of protein subunits may pass across the entire membrane (transmembrane proteins). Such traversing proteins could be in contact with the aqueous solvent on both sides of the membrane. A similar model has been postulated for the cholinergic receptor. This model also reflects the generally accepted view that the major proportion of the phospholipids in the membrane are arranged in a bilayer form.

This mosaic model of the membrane, with intercalating proteins, comes from the use of freeze-fracturing techniques in erythrocytes and other cell membranes. The red cell ghosts show a large number of particles, about 8 nm in diameter, that have been interpreted as representing proteins embedded within the plane of cleavage which passes through the middle of the lipid bilayer. There are between 500,000 and 600,000 such particles in a cell and more are attached to the inner half of the membrane (PF) than to the outer half (EF). We mentioned earlier that these particles represent dimers of the polypeptide found in *band* 3, and that they may also correspond to anionic permeation channels. The degree of dispersion or aggregation of these particles within the membrane appears to be influenced by the *spectrin-actin system.* This suggests that some link between this system and the dimers of band 3 could exist in the living membrane. The mosaic model allows for the characteristic *asymmetry* of the membrane, in which specific components predominate in the outer half or the inner half of the membrane. This implies that there are constraints on the transmembrane rotation of those macromolecules that are well oriented for the carrying of information across the cell membrane. In this model the fluidity of the membrane also implies that both the lipids and the proteins have considerable freedom of movement within the bilayer.

The fluidity of the lipids depends:

- On the degree of saturation of the hydrocarbon chains
- On the ambient temperature. A considerable proportion of the lipids in the membrane are unsaturated, so that the melting point of the bilayer is below body temperature.

Membrane Fluidity

For studying purpose, it can be classified as physical and biological.

The physical techniques are of two main classes:

1. Those that involve a minimal perturbation of the membrane, such as *X-ray diffraction* and *nuclear magnetic resonance (NMR) spectroscopy,*
2. Those that use certain added molecules to monitor specific sites of the membrane. Into this second class fall *fluorescence microscopy,* which uses fluorescent probes and *electron spin resonance (ESR)* spectroscopy.

In practice, these physical techniques can be subdivided on the basis of their ability to examine mainly either the lipid or the protein components, or both. ESR spectroscopy uses paramagnetic probes (*e.g.*, nitroxide-containing amphipathic molecules) which are introduced into the lipid bilayer. This can be done with natural and artificial membranes The data obtained provide information on lipid-protein packing, lateral diffusion of lipids, lipid-protein interactions, the fluidity of the membrane, and the rate of transmembrane rotation (so-called flip-flop) of molecules across the bilayer.

The *biological techniques* involve light and fluorescence microscopy and electron microscopy, including freeze-fracturing and radioisotope labeling methods. One of the simplest methods consists of binding gold or carbon particles to the cell surface and observing under the light microscope the movement of those particles on the surface.

More information has come from studies in which different ligands, such as antibodies and plant lectins, interact with cell surface receptors. If these ligands are labeled with fluorescent dyes their movement can be followed by fluorescence microscopy.

For example, in a classical experiment by Frye and Edidin mouse L cells and human transformed cells which have different surface antigens, were marked with the corresponding antibodies labeled with two distinct fluorescent dyes. These cells were then induced to fuse under the influence of Sendai viruses. While at the beginning both cell surfaces could be recognized by their differing labels, after 40 minutes considerable intermixing of the antigens had occurred, so that the two labels could no longer be recognized. This intermixing was retarded by temperatures below 20° C.

In surface antigens of cultured muscle cells, a diffusion rate of 1 x 10^{-9} cm per second was calculated. Also, the so-called clustering or capping effect observed in lymphocytes may result in the displacement of antigenic molecules within the surface membrane. In this case, upon treatment with a labeled antibody, the antigens are first randomly distributed, but after some time they become clustered, forming patches, and agglomerate at one pole of the cell (*i.e.*, capping) where pinocytosis of the antigen-antibody complexes takes place. This process is also inhibited by temperatures that produce solidification of the lipid bilayer.

A number of histochemical techniques using the electron microscope may give information about the fluidity of the membrane. For example, antigens labeled with ferritin or peroxidase can be used. Another extremely valuable tool is freeze-fracturing with which it is possible to demonstrate the movement of the particles in response to different experimental conditions that produce changes in the molecular structure of the membrane.

Membrane Fluidity

The fluid mosaic model has been used to explain the fact that many receptors present in the cell membrane are coupled to enzymes such as adenylate and guanylate cyclase. We can say that they are macromolecules, in most cases of a protein nature, that act at the interface between the cell surface and the environment.

A receptor has the dual function of recognizing a chemical signal (*i.e.*, one of the many regulatory agents, such as peptide hormones, neurotransmitters, prostaglandins, antigens plant lectins, and some bacterial toxins) and of initiating a biological response. The function of recognition is determined by the presence of a site that binds specifically to the ligand (*i.e.*, to the binding portion of a particular regulatory agent). This interaction produces a conformational change that may induce the translocation of ions or the activation of membrane enzymes, such as adenylate cyclase, thus producing cyclic AMP (cAMP).

In a cell there may be several kinds of receptors (at least eight in a fat cell) coupled to the same adenylate cyclase. The ligands do not compete for each specific receptor when binding, and the effect of multiple ligand-receptor bindings on adenylate cyclase is not additive. These findings have been interpreted as indicating that each ligand reacts with a unique receptor and that several receptors may be coupled to a certain number of enzyme molecules in the membrane.

According to the *mobile hypothesis* both the receptors and the enzymes can diffuse independently within the plane of the membrane The greater degree of freedom that this model includes within the lipid realm of the membrane would allow different receptors to couple to a single enzyme.

Fusing cells in which the adenylate cyclase was inactivated but the receptor remained intact, with other cells which contain only the enzyme, resulted in receptor-enzyme coupling. In other words, the receptor of one cell (*e.g.*, the b-adrenergic receptor of an erythrocyte) was coupled to the adenylate cyclase of the other cell (*e.g.*, a tumor cell lacking the receptor). The coupling is manifested by a great increase in cAMP after receptor interaction with the specific ligand (in this case isoproterenol).

The diagram shown below interprets in a highly simplified manner the fusion of two different cells that results in coupling between a receptor from one cell and adenylate cyclase from the other.

The Myelin Sheath and the Photoreceptors

Important information about the molecular structure of the cell membrane is derived from studies of natural multi-layered lipoprotein systems, such as the myelin sheath and the photoreceptors.

The Myelin Sheath: The myelin sheath is lipoprotein membrane that surrounds the axon of the nerve fibre. In peripheral nerves this sheath is formed by the Schwann cells. In central nerves the myelin sheath is produced by the activity of the oligodendroglial cells. It is a very special membrane structure in which the lipids are more abundant than the protein. In central myelin there are two main protein moieties: the proteolipid and the basic protein.

The Myelin Sheath has been known for over a century that the myelin sheath has a strong birefringence, which indicates a high degree of organization at a submicroscopic level. Studies with x-ray diffraction have revealed a spacing of 17 nm in amphibian and 18.0 to 18.5 nm in mammalian peripheral nerves. Within this period, the proportion corresponding to the lipid, protein, and water content has been estimated.

The existence of lecithin-cholesterol and Sphingomyelin-cholesterol complexes is postulated. The first type can be accommodated within the thickness of the lipid layer (L), but in the second, the longer sphingomyelin molecules must interdigitate in order to fill the same space. This model also accounts for the localization of protein (HP) and water (HL) and is in accord with the view that each x-ray diffraction period of 18 nm corresponds to two unit membranes, *i.e.*, to two lipid bilayers separated by a watery space (H), that originally corresponds to extracellular space.

It has confirmed by electron microscopic studies that myelin has a multi-layered membranous structure. In most cases, however, the x-ray diffraction period is reduced to 10 to 12 nm, thereby introducing a great many artifacts. Recently, with a highly polar embedding medium based on polymerized glutaraldehyde-urea, the periodicity obtained was of the same order as by x-ray diffraction. With this technique the lipids are not extracted; this explains the lacks of collapse of the structure.

In nerve conduction to function as an insulator, preventing the dissipation of energy into the surrounding medium. It might act not only as a dielectrics (insulating) material but also as a kind of resonant conductor in which the energy waves resonating in the lipid layers between the protein membranes could pass with maximum speed and minimum loss of energy.

Photoreceptors

The retinal rods and cones are highly differentiated cells that have at their outermost segment a lipoprotein structure that is specialized for photo-reception. Studies with the polarization microscope suggest a submicroscopic organization consisting of transversely oriented protein layers alternating with lipid molecules arranged longitudinally along the axis of the photoreceptor. This type of layered organization has been demonstrated by electron microscopy in fragmented rod outer segments and in thin sections of the retina. These observations indicate that the rod consists of a stack of superimposed disks (several hundred) along the axis. These disks are really flattened sacs made of two membranes, which surround a thin space of 3 nm and become continuous at the edges. The space between the rod sacs is 5 to 12 nm. The cone outer segments, with minor differences, have a similar structure.

To osmotic change rod sacs are highly sensitive. In hypotonic solutions they swell considerably, and the inner space between the membranes becomes very large. The rod outer segments can easily be isolated in the frog and be submitted to chemical analysis. The prótein composition is simpler than in other membranes. The photopigment protein, *rhodopsin,* represents 40 per cent of the total mass of the rod sacs; another 40 per cent is lipid. Rhodopsin is an integral type of protein requiring detergents for extraction. In the pline of the membrane the rhodopsin molecules have a liquid-like distribution. X-ray diffraction studies have led to the interesting conclusion that this molecule may change its position within the lipid layer. In the dark (non-stimulated condition) the molecule is about one third embedded in the lipid and two thirds exposed to the water surface. Upon activation by light (bleached rhodopsin) the molecule becomes more deeply embedded within the lipid bilayer. This change is presumably caused by changes in surface charges of the molecule.

The process of photoreceptors transform light energy into another type of energy that can be conducted as nerve impulse.

It is based on a cycle of chemical reactions which involve the visual pigments present in the protein membranes of the rod and cone sacs. This multilayered structure is a very effective system that facilitates the maximum absorption and utilization of light by the chromophoric groups present in the visual pigments (retinenes). The acute sensitivity of the photoreceptors, which can react to a single photon, can be explained by the fact that the possibility of striking a sensitive molecule is increased by a factor of hundreds or thousands by the molecular organization of the photoreceptor.

It is known that the interaction with photons of light produces liberation of Ca^{2+} ions which, in turn, inhibit the sodium current that is characteristic of the photoreceptor in the dark condition. In this way photoreceptors act as energy transducers, transforming light energy into electrical signals.

CELL PERMEABILITY

For functioning of the living cell and maintaining of satisfactory intercular physiologic conditions permeability is fundamental This function determines which substances can enter the cell; many of which may be necessary to maintain its vital processes and the synthesis of living substances. It also regulates the outflow of excretory material and water from the cell. The presence of a membrane establishes a net difference between the *intracellular* fluid and the *extracellular* fluid in which the cell is bathed This may be fresh or salt water in unicellular organisms grown in ponds or the sea, but in multicellular organisms the internal fluid, *i.e.*, the blood, the lymph, and especially the *interstitial* fluid, is in contact with the outer surface of the cell membrane one of the functions of the cell membrane is to maintain a balance between the osmotic pressure of the intracellular fluid and that of the interstitial fluid.

In higher organisms, the osmotic pressure of the body as a, whole is regulated principally by the kidneys, and the osmotic pressure of the interstitial fluid is about the same as that of the intracellular fluid. In plants, the intracellular fluid has a higher osmotic pressure than the extracellular fluid. The cell is protected from bursting by a rigid cellulose wall. Animal cells generally lack the turgidity that characterizes plant cells. The unfertilized eggs of some marine animals, such as the sea urchin, behave like genuine osmometers. Since they are spheroid, one can, by measuring the diameter, determine the volume and the changes that the egg undergoes with changes in the osmotic pressure of the medium.

The osmotic equilibrium is maintained by means of a contractile vacuole, in many unicellular. This "organelle" extracts water from the protoplasm, releasing its contents into the external medium.

Electrical Potentials acroes

In all cells there is a difference in ionic concentration with the extracellular medium and an electrical potential across the membrane. These two properties are intimately related, since the electrical potential depends on an unequal distribution of the ions on both sides of the membrane. The interstitial fluid has a high concentration of Na^+ and Cl^-; and the intracellular fluid, a high concentration of K^+ and of larger organic anions (A^-). Using fine microelectrodes with a tip of 1 mm or less, investigators are able to penetrate through the membrane into a cell and also into the cell nucleus and to detect an *electrical potential* (also called the *resting,* or *steady,* potential), which is always negative inside. The values of the membrane potential vary in different tissues between -20 and -100 millivolts or mv.

Passive Permeability

The passage of molecules and ions across membranes may be the result of two principal mechanisms. Permeability may be *passive,* if it occurs only because of physical laws such as *diffusion* or *active* if it requires energy for transport. There are several kinds of permeability which will be considered in the following sections. Without an intervening membrane, when two solutions of different concentration are mixed a process of intermixing called *diffusion* occurs. For example, if a concentrated solution of sugar is placed in contact with water, there will be a net movement (*i.e.,* flux) of the solute from the region of higher concentration to that of a lower concentration. In this case the higher the difference in concentration between the two solutions (*i.e.*, the *concentration gradient*), the more rapid the rate of diffusion.

The presence of a lipoprotein membrane, such is the plasma membrane, greatly modifies this diffusion or passive permeability. Overton demonstrated that substances that dissolve in lipids pass more easily into the cell, and Collander and Barlund, in their classic experiments with the cells of the plant *Chara*, demonstrated that the rate at which substances penetrate depends on their solubility in lipids and the size of the molecule. The more soluble they are, the more rapidly they penetrate, and with equal solubility in lipids the smaller molecules penetrate at a faster rate.

The permeability (P) of molecules across the membrane is:

$$P = \frac{KD}{t}$$

with K, the partition coefficient; D, the diffusion coefficient, which depends on the molecular weight; and t, the thickness of the membrane The partition coefficient in most cell membranes is similar to that of olive oil and water.

The *partition coefficient* can be measured by mixing the solute with an oil-water mixture and then waiting until the phases are separated. The coefficient is the concentration of the solute in the oil, divided by the concentration in the aqueous phase. Today, the *diffusion coefficient* can be easily determined by using radioactive solutes and measuring their rate of entry into the cytoplasm at various external concentrations. From the experiments. It is evident that many molecules may enter the cell by simple diffusion if the external concentration is higher than the internal. On the other hand, waste products accumulated in the cytoplasm may be eliminated by diffusion.

Passive Ionic Diffusion-Dependent on the Concentration and Electrical Gradients

The diffusion of ions across membranes is even more difficult because the same is depends not only on the concentration gradient but also on the electrical gradient present in the system. Within the cell there is a large concentration of non-diffusible anions (A^-) and that an electrical gradient across the membrane is established.

Donnan predicted in 1911 that if a theoretical cell, having a non-diffusible negative charge inside, is put in a solution of KC1, K^+ will be driven into the cell by both the concentration and the electrical gradients; Cl^-, on the other hand, will be driven inside by the concentration gradient, but will be repelled by the electrical gradient. As shown by Donnan, the equilibrium concentrations will be exactly reciprocal:

$$\frac{[K^+{}_{in}]}{[K^+{}_{out}]} = \frac{[Cl^-{}_{out}]}{[Cl^-{}_{in}]}$$

According to Donnan equilibrium involving only physical forces (*i.e.*, without expenditure of energy by the membrane) was confirmed by the demonstration that the membrane potential was negative on the inside and was accompanied by a high K^+ and a low Cl^- concentration.

The relationship between the concentration gradient and the resting membrane potential is given by the Nernst equation:

$$E = RT \log \frac{C_1}{C_2}$$

where E is given in millivolts, R is the universal gas constant, and T is the absolute temperature.

Donnan equilibrium for KG can now be expressed as follows:

$$E = RT \log \frac{[K^+{}_{in}]}{[K^+{}_{out}]} = RT \log \frac{[Cl^-{}_{out}]}{[Cl^-{}_{in}]}$$

Any increase in the membrane potential will cause an increase in the ion asymmetry across the membrane, and vice versa. While the first measurements of membrane potentials and ion concentration seemed to confirm this type of *passive* or *diffusion equilibrium*, more precise determinations in different cell types demonstrated that this was not the case. As mentioned in the next section, this discrepancy may be explained by the involvement of the active transport of ions.

Active Transport of Ions

Extra than the diffusion movement of neutral molecules and ions across membranes, cell permeability includes a series of mechanisms that require energy. These medianisms are generally described as *active transport*. Adenosine triphosphate (ATP), which is produced mainly by oxidative phosphorylation in mitochondria, is generally used as the source of energy. For this reason, active transport is generally related to, or coupled with cell respiration.

There is explained active transport against a concentration gradient in which water has to be moved upstream (*i.e.*, against gravity). The osmotic work to be done is expressed by the Nernst equation, ft charged molecule crossing through an electrochemical gradient also may imply expenditure of energy. For example, to maintain a low intracellular concentration (Na^+, the cell must extrude sodium against a gradient (*i.e.*, higher Na^+ concentration outside). In addition, it must do this against an electrochemical barrier since the membrane is negative inside and positive outside.

A Sodium Pump

During transportation of ion against an electrochemical gradient an extra consumption of oxygen is required. It is calculated that 10 per cent of the resting metabolism of a frog muscle is used for transport of sodium ions. This consumption may increase to 50 per cent in some experimental conditions in which the muscle is stimulated.

The resting membrane potential is due to active transport may be demonstrated in plant and animal cells that have been metabolically blocked by anoxia or specific poisons. In this case leakage of K^+ occurs and the potential may decrease to zero. This is clearly observed when anoxia is combined with the poisoning of glycolysis. As suggested by Krogh in 1946. The membrane potential is not really at equilibrium but in a "steady state" involving the constant expenditure of energy.

There is a example of active transport has been provided by experiments with isolated frog skin. The epithelium is specialized to transport Na^+ from the pond water to the interstitial fluids, and by this mechanism the frog can trap this essential ion for use in different tissues. The isolated skin can be kept alive for many hours and used as a wall between two chambers in which the ionic concentration and other factors are changed experimentally.

By means of this preparation, a difference in potential across the skin has been demonstrated: the inside surface is positive with respect to the outside. The sodium ions are transported from the outer towards the inner surface, and the current produced is due to the flux of sodium. At was also observed that the antidiuretic hormone of the neurohypophysis stimulates the transport of sodium and water. Similar findings have been observed in the isolated toad bladder and the kidney tubules of *Necturus*.

Ionic transport is intense in the salivary and sweat-producing glands, and even more so in the glands of the stomach that produce H^+ and $C1^-$ that must be replaced by the blood. Transport of ions is fundamental to maintenance of the osmotic equilibrium of the cell, the required concentration of anions and cations and the special ions needed for the functioning of the cell. Together with the extrusion of Na^+, which is continuously pumped out by the cell, there is an exit of water molecules. In this way the cell keeps its osmotic pressure constant.

Potassium ions, which are concentrated inside the cell, must pass against a concentration gradient. This can be achieved by a "pumping" mechanism at the expense of energy. As explained earlier, Na^+ also is transported by an active process, which is sometimes called the "sodium pump." The passive (downhill) fluxes are distinguished from the active (uphill) fluxes. Notice that the active pumping out of Na^+ is the main mechanism for maintaining a negative potential inside the membrane of -50 mv. This diagram demonstrates that the distribution of ions across the membrane depends on the summation of two distinct processes:

1. Simple electrochemical diffusion forces which tend to establish a Donnan equilibrium (*i.e.*, passive transport)

2. Energy-dependent ion transport processes (*i.e.*, active transport)

It may be observed that the Na^+— K^+ pump drives ions extruding Na^+ and taking in K^+ (active fluxes). At the same time the passive Na^+ influx depends on a large driving force (D.F.), resulting from the concentration and voltage gradients, and only a slight permeability of the membrane to Na^+ (P_{Na}). Similarly, the net passive K^+ efflux results not from a small driving force, but from a greater permeability to K^+ (P_k).

Ionic Transport through Charge Pores In the Membrane

The molecular machinery involved in ionic transport is located within the cell membrane. This has been demonstrated in two key materials. For example, if red blood cells are hemolyzed so that only the cell membrane remains, they can be filled again with appropriate solutions containing ions and ATP, and Na^+ is transported and K^+ is taken up as in a normal cell.

The giant axon of the squid, which has a diameter of about 0.5 mm, can be emptied of the axoplasm and then refilled with solutions of different electrolytes. The transport of ions against a concentration gradient, steady potentials, and even action potentials with the conduction of impulses can be obtained in this preparation in which most of the axoplasm is lacking and the excitable membrane left alone.

The utilization of radioscopes demonstrated that ions can enter into the cell rapidly without obvious osmotic effects. It was then suggested that an ionic interchange across the membrane could take place through electrically charged pores. Knowledge about the diameter of the different ions in the hydrated state is particularly pertinent. In this respect it is interesting to remember that the sodium ion, although smaller than K^+ and Cl^- in weight, is large in the hydrated condition and enters with more difficulty into the cell.

The pore could be envisioned as the interstice between four adjacent protein subunits which could form a hydrophilic channel across the membrane. The ion channels for Na^+ and K^+ are represented in the polarized or resting condition, or in the state of depolarization that occurs during the action potential. It is postulated that the Na^+ and K^+ channels are closed in the polarized resting membrane.

The rest of the diagram shows the opening of the Na^+ pores at the peak of depolarization of the action potential and the opening of the K^+ pores during repolarization. In red blood cell the total area of the pores has been estimated to be on the order of 0.06 per cent of the surface area. This means that a 0.7 nm pore would be surrounded by a nonporous square 20 x 20 nm. These findings indicate that the cell uses only a minute fraction of its surface area for ionic interchange.

Anion Transport in Erythrocytes

From tissues to the lung the red cell membrane is endowed with an extremely active anion permeation mechanism to transport CO_2 This system can exchange chloride and bicarbonate ions across the membrane. Using several chemical probes that inhibit anion permeability, knowledge has been gained about the molecular localization of this mechanism. Some of the probes can be transported by the anion system and, under certain conditions, can be fixed covalently to it. In this way it is possible to determine not only the number of permeation sites but also the polypeptide in which they are localized. This approach has permitted the identification of the *polypeptide of band* 3 with the site involved in the transport. This protein spans the membrane and may exist as a dimer (or tetramer). We have seen earlier that it also corresponds to the 8 nm particles observed in freezefractured membranes.

The proposed model for anion transport is that of a continuous proteinaceous aqueous channel across the membrane having, near the outer surface, one anion binding site with three positive charges, and a hydrophobic barrier to limit the free diffusion of anions. Passage through the barrier could occur only as a consequence of the binding of the anion.

In the above model it is assumed that the segment containing the binding site could exist in two conformations, one facing outward, the other inward. In this way the segment could act as a gate in the proteinaceous channel, swinging between the two positions; this would permit the binding site to interact with anions coming from the outside or from the interior of the cell.

In this model it is observed that the anion binding site is located in a 65,000 daltons segment of the protein which remains in the membrane after a 35,000 daltons segment has been separated by a proteolytic enzyme.

The Vectorial Function of N^+ K^+ *ATPase*

Previously we have studied that the so-called *sodium pump* is a mechanism of active transport by which Na^+ is eliminated from the cell. This mechanism, discovered by Hodkin and Keynes in 1955, was soon associated by Skou with the Na^+ K^+ ATPase. This enzyme is able to couple the hydrolysis of ATP with the removal of Na^+ from the cytoplasm against an unfavourable electrochemical gradient.

It may be observed that the hydrolysis of one ATP provides the energy for the linked transport of 2 K^+ ions towards the inside, and 3 Na^+ ions towards the outside of the cell.

This diagram shows the vectorial characteristics of the enzyme which is sensitive to ATP on the inside of the membrane, but not on the outside. This ATPase is stimulated by a mixture of both Na^+ and K^+ and is inhibited specifically by the cardiotonic glycoside *ouabain.* The Na^+ and K^+ sites are independent and are competitively inhibited by K^+ and Na^+, respectively.

The Na^+ K^+ ATPase has been found to be concentrated in membranes of nerves, brain, and kidney and is particularly rich in the electric organ of the eel and the salt gland of certain marine birds (*e.g.*, albatross) in which the transport of ions is extremely active. The vectorial properties of the enzyme can be demonstrated cytochemically; thus, by electron microscopy the liberation of phosphate has been shown to occur exclusively on the inner side of the red cell ghost. On the other hand, ouabain binds exclusively to the outer surface of the membrane. The vectorial function of Na^+K^+ ATPase can be best explained by the so-called "carrier" hypothesis.

In its most simple form a carrier consists of:

- A membrane protein having a specific site to combine with the ligand to be translocated.
- A mechanism of translocation Which moves the ligand from one side of the membrane to the other.
- A mechanism for the release of the ligand. Most carrier models involve some kind of conformational change of the protein by which the affinity of the binding site for the ligand changes during the translocation.

In the case of Na^+ K^+ ATPase it is possible to envision that the enzyme has binding sites for Na^+ and K^+ and a complex carrier mechanism to deliver and release these ions at opposite sides of the membrane. The coupling of the barrier mechanism to the hydrolysis of ATP generates the energy needed to move the ions against the concentration gradients. There are two steps of transportation.

Step. I consists of the formation of a *covalent phosphoenzyme intermediate* (E ~ P) on the inner side of the membrane and in the presence of Na^+. This first reaction is inhibited by Ca^{2+}, but not by ouabain. In Step II the [Na^+ E ~ P] complex is hydrolyzed to form free enzyme and Pi. This reaction requires K^+, added to the outside of the membrane, and is inhibited by ouabain, which competes with K^+.

Steps I and II can be represented as given below:

Step I $Na^+_{in} + ATP + E \longrightarrow$

$[Na^+ \cdot E \sim P] + ADP$

Step II $[Na^+ \cdot E \sim P] + K^+_{out} \longrightarrow$

$Na^+_{out} + K^+_{in} + Pi + E$

Na^+ and K^+ are ultimately translocated in the opposite directions from which they were bound. The Na^+ K^+ ATPase is tightly bound to the cell membrane, from which it may be released by the use of detergents. The molecular weight of the enzyme has been estimated to be about 670,000, and it probably contains several polypeptides. Certain lipids are needed for the activity of this enzyme. In a red cell there are about 5000 enzyme molecules, each of which may extrude 20 Na^+ ions per second.

Transport of Various Substances

There are two types of Na^+ pumps which can be distinguished:

1. The so-called *Na^+ K^+ exchange pump or coupled neutral pump* in which, as described before, the transport of K^+ inward is coupled with that of Na^+ outward and in which electrical neutrality is thus achieved.
2. The so-called *electrogenic pump* in which the exit of Na^+ is not compensated by the entrance of K^+. In this case an electric gradient generates the potential which may provide the driving force for transporting other solutes.

The transport of many substances, such as glucose and various amino acids, into the cell may be coupled to the Na^+ pump. For example, the entrance of glucose into the intestine cells is activated by Na^+ and inhibited by K^+ and ouabain. To explain these findings a carrier mechanism has been postulated. In such a model the carrier should have a binding site for Na^+ and another for glucose and should exist in two conformational states, one with low affinity for glucose and the other with high affinity.

In postulated model in which the glucose site is first in a high affinity state and binds the glucose molecule at the surface of the cell. In a second step the carrier is translocated, and the glucose site changes to a low affinity and releases the ligand into the cell. The two affinities are modulated by the concentration of Na^+ outside and inside the cell.

The carrier model is based on the existence of a protein that is able to translocated a soluble molecule across the thickness of the membrane. Several hypotheses for the molecular mechanism that is involved have been put forth. In one, the solute binds on the outside surface of the membrane to the carrier, and this then rotates and delivers the solute to the interior of the cell. In the second, it is supposed that the carrier is fixed in place but is able to undergo some kind of conformational change by which the solute is translocated. From the viewpoint of energetics this *fixed-pore mechanism* is more probable, since it would require less expenditure of energy.

Selectivity of Transport

Transportation in different molecules across the membrane shows a high degree of specificity. In other words, the permeability of a molecule is related to its chemical structure. For example, while one molecule may readily enter the cell, another, with the same size but a slightly different molecular structure, may be completely excluded. This type of selectivity is similar to that of the *enzymes* and to that of the *receptors,* in which a *binding site* is able to "recognize" a definite *substrate* or *ligand.* Such, selective transport has been attributed to certain proteins of the membrane which are called *permeases.*

Permeases *accelerate transport,* provide for *special selectivity,* and are *recycled;* this means that they remain unchanged after having assisted in the entry or exit of a molecule. Some permeases can transport only if there is a favourable concentration gradient, while others may also do so against an unfavourable one. The first type of transport, being driven by a passive mechanism, is sometimes called *facilitated diffusion.*

Incase the permease transport is against the gradient, it represents another case of *active transport.* One of the best known permease systems is the *lac permease of E. coli.* One lac permease is coded within the *lac operon* and is a single unit protein that acts as a carrier for the entrance of lactose.

Permeases are genetically determined; for example, in *E. coli* there are mutants which are incapable of transporting lactose. Lac-permease has been isolated and purified. In E. coli there may be many permease systems specialized for the transport of various molecules. Permeases are found in the cell membrane of eukaiyotic cells and also in the inner mitochondrial membrane.

Penetration by Large Molecules—Various Mechanisms

What has been said demonstrates that the general term "cell permeability" comprises a variety of different mechanisms. In addition to certain foreign substances that can penetrate the cell because they are lipid-soluble (*e.g.,* anesthetics), generally ions penetrate through charged pores and other molecules penetrate by permease systems.

There is no doubt that under certain conditions large molecules, such as certain proteins, penetrate the cell. This is the case with ribonuclease, an enzyme that penetrates living plant cells readily and also eggs, flagellates, ascitic tumors, and so forth. Basic proteins of the protamine and histone types have been reported to enter into living cells.

DIFFERENTIATIONS AT THE CELL SURFACE AND INTERCELLULAR COMMUNICATIONS

Regions of the cell surface of certain cells are related to absorption, secretion, fluid transport, and other physiologic processes. Topographically they are referred to as specializations of the cell surface. The apical surface is projected into slender processes called *microvilli.* On the surface corresponding to the edge of the cell in contact with an adjacent cell there are several differentiations. At the base of the cell the plasma membrane is covered by a thick basement membrane of extracellular material where infoldings of the plasma membrane may be known.

Microvilli—A Greatly Increased Cell Membrane Surface Area

Microvilli are very prominent and form a compact structure in the intestinal epithelium that appears under the light microscope as a *striated border*. These microvilli, which are 0.6 to 0.8 mm long and 0.1 mm in diameter, represent cytoplasmic processes covered by the plasma membrane. Within the cytoplasmic core fine microfilaments are observed which in the subjacent cytoplasm form a terminal web. As these filaments contain actin and are attached to the tips of the microvilli by a-actinin; their function is to produce contraction of the microvilli.

The outer surface of the microvilli is covered by a coat of filamentous material (*fuzzy coat*) composed of glycoprotein macromolecules. Microvilli increase the effective surface of absorption. For example, a single cell may have as many as 3000 microvilli, and in a square millimeter of intestine there may be 200,000,000. The narrow spaces between the microvilli form a kind of sieve through which substances must pass during absorption. Numerous other cells have microvilli, although they are fewer in number. They have been found in mesothelial cells, in the epithelial cells of the gall-bladder, uterus, and yolk sac, in hepatic cells, and so forth.

The *brush border* of the kidney tubule is similar to the striated border, although it is of larger dimensions. An amorphous substance between the microvilli gives a periodic acid Schiff reaction for polysaccharides. Between the microvilli the cell membrane invaginates into the apical cytoplasm. These invaginations are apparently pathways by which large quantities of fluid enter by a process of pinocytosis.

Desmosomes and Intermediary and Tight Junctions

There are four types of differentiations are present at the lateral surfaces of epithelial cells:

- The macula adherens or desmosome;
- The zonula adherens, also called intermediary junction;
- The zonula occludens or tight junction;
- The gap junction or nexus.

In a number of epithelial cells, the desmosomes, are found which appear under the light microscope as darkly stained bodies. These structures are formed by a circular area, about 0.5 mm in diameter, of the plasma membranes of two adjacent cells that are separated by a distance of 30 to 50 nm.

Under the membrane there is a dense intracellular plaque towards which numerous *tonofilaments* converge. These filaments describe a kind of loop in a wide arc and course back into the cell. Within the intercellular gap a coating material may be observed which sometimes forms a discontinuous middle dense line. This extracellular material contains acid mucopolysaccharides and proteins. In fact, desmosomes are broken by trypsin, collagenase, and hyaluronidase and are sensitive to agents that chelate calcium. The dense intracellular attachment plaques are digested by proteolytic enzymes.

When the tonofilaments provide the intracellular mechanical support, cellular adhesion at the desmosome depends on the extracellular coating material. Frequently there are regions of looser contact between the desmosomes and even intercellular spaces for free circulation of fluids. Along the basal surface of some epithelial cells *hemidesmosomes* may be observed. These are similar to desmosomes in fine structure, but represent only half of them, the outer sides frequently being substituted with collagen fibrils.

The *terminal bar* also called *intermediary junction* is generally found at the interface between columnar cells just below the free surface. Under the electron microscope the terminal bar appears somewhat similar to the desmosome. The membrane is thickened and the adjacent material is dense, but filaments are generally lacking. Within the *zonula adherens* there are parts of the two adjacent membranes that come into close contact at certain points, forming a kind of anastomosing network that extends around the apical region of the cell. Such a network is especially visible in freeze-etched preparations.

Cell contacts may be specially differentiated to create a barrier or "seal" to diffusion. In the *tight junction (zonula occludens)* the adjacent cell membranes adhere, and therefore there is no intercellular space for a variable distance. The tight junction is situated just below the apical border, and at this point the outer leaflets of the unit membranes adhere, forming a single intermediary line. Experiments have demonstrated that macromolecules that are put into the lumen of an epithelial cavity cannot penetrate the intercellular space. Tight junctions may also play an important role in brain permeability at the level of the blood-brain barrier and the synaptic barrier.

Gap *Junctions Intercellular Communications*

As the basic unit of all living matter cell concept should not lead us to disregard the fact that multicellular organisms are made of populations of cells that interact among themselves. Such *cellular interaction* is essential for the coordination of activities, and furthermore, the propagation between cells of signals for growth and differentiation is indispensable for development.

It is now known that most cells in an organized tissue are interconnected by *junctional channels* and that they share a common pool of many small metabolites and ions that pass freely from one cell to another. However, their individuality is maintained by macromolecules which are not exchanged between cells.

Electrical Coupling between Cells Depends on Gap Junctions

By introducing intracellular microelectrodes into adjacent cells of a tissue, investigators have been able to demonstrate that in many animal cells there are intercellular communications. In this case the cells are electrically coupled fluid have regions of low resistance in the membrane through which there is a rather free flow of electrical current carried by ions.

The other parts of the cell membrane, which are not coupled, show a much higher resistance. This type of coupling, called *junctional communication,* is found extensively in embryonic cells. In adult tissue it is usually found in epithelia, cardiac cells, and liver cells. Skeletal muscle and most neurons do not show electrical coupling. An excellent material for the study of gap junctions is the myocardial tissue, in which the action potential is transmitted from cell to cell by an electrical coupling.

In a thin section the gap junction appears as a plaque-like contact in which the plasma membranes of adjacent cells are in close apposition, separated by a space of only 2 to 4 nm. This gap can be filled with electron opaque materials such as lanthanum and ruthenium red. In tangential sections gap junctions show a hexagonal array of 8- to 9-nm particles. The electron-dense material is able to penetrate in between the particles, thus delineating their polygonal arrangement, and also into the central region of each particle. This central region is 1.5 to 2 nm in diameter and corresponds to the probably location of the channel.

With freeze-fracturing it is possible to split the junctions and to better define their internal structure. Upon the inner half-membrane of the fracture (PF face) the 8-to-9-nm particles are observed, while in the external half (EF face) a complementary array of pits or depressions appears. Gap junctions are resistant to mechanical disruption, proteolysis, and removal of Ca^{2+} and can be isolated from the cells. After isolation they can be shown by electron microscopy and x-ray diffraction to have the same polygonal lattice structure, with 8-to-9—nm particles, as seen in the intact junction The unit of the gap junction has been named the *"connexon,"* and it appears to span the bilayer, of each of the two connected cell membranes, as well as the gap in between, thus providing for intercellular communication.

The development of gap junctions has been observed in various cell types. The process consists of the appearance of plaques between cells, with a reduction in the intercellular space, the appearance of larger particles (probably precursors), and finally the arrangement of the particles in polygonal arrays.

When two cells are brought into direct contact, junctional communications may be formed in a matter of seconds, and this does not require new protein synthesis. It is therefore thought that the 8 nm protein particles are already present on the cell surface and that when cells come in contact these particles diffuse and interact with similar particles on the other membrane.

From the physiologic viewpoint we can envision the gap junction as resulting from the apposition of two channels, one present in each membrane. Such channels not only traverse each membrane but also project into the intercellular gap and are joined there, thus creating the direct intercellular communication of the connexon. It can be observed that the channels represent a region of high permeability that is separated from the intercellular fluid by a kind of junctional seal of low permeability.

Gap Junctions

The size of the channels present in gap junctions has been estimated by:

- Electron microscopy
- By electrical measurements, and
- By the diffusion of tracer substances.

By determining the conductance of single channels and the resistance of the cytoplasm and by assuming a cylindrical channel 20 nm long, the diameter can be calculated to be of the order of 1 nm, a size which is in agreement with morphological data.

The first tracer used was the fluorescent dye *fluorescein* (300 daltons), which when injected into a cell was shown to diffuse into adjacent cells. Recently, with the use of fluorescent peptides in cells of *Chironomus* salivary glands) a weight limitation of 1300 to 1900 daltons was reached, which corresponds to a channel pore of about 1.4 nm.

The permeability of the channel is regulated by Ca^{2+}. Normally the Ca^{2+} content of the cell is low (10^{-7} M), and the permeability of the channel is high. If the intracellular Ca^{2+} concentration increases; the permeability is reduced. The extracellular concentration of Ca^{2+} is about 10^{-3} M, so that any damage to the cell membrane or the junctional seals may increase the Ca^{2+} inside the cells. When the Ca^{2+} level reaches a certain limit the channels close, and they open again if the cell eliminates the excess Ca^{2+}.

It has been known that the junctional communications depend on the energy provided by oxidative phosphorylation (*i.e.*, ATP). Treatments that inhibit cell metabolism, such as cooling at 8° C, dinitrophenol, cyanide, and, oligomycin, produce uncoupling between the cells; this may be reversed by injection of ATP. It is thought that the uncoupling may be due to the influx of Ca^{2+}, either from the medium or from mitochondria, during the action of the inhibitors.

We have studied earlier that the intracellular Ca^{2+} concentration has been monitored by the injection of *aequorin*, a luminescent protein in which the light emission is proportional to the Ca^{2+} concentration. With such a monitoring system and with the use of fluorescent peptides of various sizes, it has been possible to demonstrate that the permeability changes caused by Ca^{2+} are *graded*. In other words, tracers of different molecular sizes can cross the channel according to the intracellular Ca^{2+} concentration. As the Ca^{2+} is elevated from 10^{-7} to 10^{-5}M, the limit of permeation decreased gradually, and at 5 ′ 10^{-5}M the channel is closed. It is thought that this control by Ca^{2+} may provide a powerful mechanism for regulation of intercellular communications.

Coupling between Cells

In cardiac muscle and in electrical synapses between certain neurons, the junctions are related to communication of electrical signals between these cells. Junctional interconnections are particularly widespread during embryonic development at stages in which cell differentiation takes place. Because diffusion of molecules can readily take place through them, these junctions are well suited for the dissemination of signals controlling cellular growth and differentiation at close range.

The coupling between cells is not only electrical but also *metabolic.* In addition to permitting free passage of electric current, junctional communications allow the passage of ions and small molecules such as nucleotides, sugars, vitamins, and other metabolites. This gives rise to the phenomenon of *metabolic cooperation* between cells.

A few cells prelabeled with a nucleotide were added to a culture of unlabeled cells; it can be seen that the nucleotide is transferred only between those cells that have direct contacts. Other experiments have shown that while nucleotides are, easily transferred, macromolecules such as enzymes and RNA are not transferred between cells.

Metabolic cooperation was first demonstrated in cell cultures of fibroblasts in which a mutation made the cells unable to incorporate exogenous ^{3}H-hypoxanthine into their nucleic acids. If such cells were grown in contact with wild-type fibroblasts (*i.e.,* non-mutants), there was incorporation in both types of cells. These findings indicate that a metabolite (the nucleotide inosine monophosphate) from the wild-type fibroblast had penetrated the mutant fibroblast through intercellular channels.

Another findings which is important certain cancer cells show no intercellular coupling and that they fail to communicate with normal cells. It is assumed that these cancer cells have a genetic defect which has interrupted the passage of growth-controlling molecules between them. Recent experiments in which heterokaryons have been produced between a cancer cell and a normal cell (*i.e.,* cell fusion) have shown that it is possible to obtain a noncancerous hybrid. In this hybrid the intercellular communications, which were absent in the original cancerous cell, were now established between the hybrid cells.

Hybrid cells were made between human normal fibroblasts and mouse cancer cells that showed no intercellular communication. These hybrids were observed for several generations by studying the type of growth (*i.e.,* cancerous or normal) and the presence or absence of channels. It has been found that a human factor probably linked to one chromosome can correct both for the growth and the channel defect.

These experiments and others suggest that coupling and permanent uncoupling could be genetically determined. They also point to the possible regulatory role of molecules of small molecular weight—probably nucleotides —between cells controlling growth and differentiation. Such regulatory agents are not transferred to the cancerous cells having no intercellular communication. It is evident that in most cells, as in a society, the exchange of information is essential to maintain normal health.

COATS OF THE CELL MEMBRANE

Regarding the coats surrounding cell membranes the same are very conspicuous in eggs of marine animals and in amphibia. A glycoprotein-like substance called *mucin* is the main constituent. Mucins also cover and protect the cell surface lining the gastrointestinal tract. Polysaccharides constitute the *pectin* and *cellulose* of plant cell walls and the *chitin of* Crustacea.

In the base of most epithelial cells glycoproteins and polysaccharides, in the form of hyaluronic acid, are found The name glycocalyx has been coined to designate the glycoprotein and polysaccharide covering that surrounds many cells.

The glycocalyx contains the oligosaccharide side-chains of the glycolipids and glycoproteins which are exposed to the outer surface of the membrane. The cell coat has negatively charged sialic acid termini, on both the glycoproteins and gangliosides, which may bind Ca^{2+} and Na^{+} ions. When the membranes are treated with *neuraminidase,* an enzyme that removes sialic acid, there is a reduction in the negative charge of the membrane. The use of PAS and Alcian blue staining may render the surface of many cell types more visible. Staining with ruthenium red or with lanthanum may reveal the thin cell coat present in biliary capillaries.

Visualization of the carbohydrates may be made more specific. These are proteins, derived from plants, that bind to the cell surface and cause agglutination. *Concanavalin A* isolated from Jack beans is specific for glucose and mannose residues; and *germ agglutinin,* for N-acetylglucosamine. These lectins can be labeled with fluorescent dyes or with electron-dense materials for observation under the electron microscope.

The cell coat generally appears as a layer 10 to 20 nm thick which is in direct contact with the outer leaflet of the plasma membrane. In some cases, as in the ameba, the coat is formed by fine filaments 5 to 8 nm thick and 100 to 200 nm long. The strength and durability of the coat varies from cell to cell. For example the coat of the intestinal epithelium resists vigourous mechanical and chemical attacks; in other cells the coat is labile and is depleted by washing or exposure to enzymes.

The biosynthesis of the glycoproteins forming the glycocalyx takes place in the ribosomes of the endoplasmic reticulum and that the final assembly with the oligosaccharide moiety is accomplished in the Golgi complex. The cell coat can be considered to be a secretion product of the cell which is incorporated into the cell surface and undergoes continuous renewal. The cell components undergo active turnover, and some of them are shed into the surrounding medium. This is particularly evident in cultured cells, which leave a kind of "footprint" on the glass to which they adhere.

Cell Coat Functions

Extra than the protection of the cell membrane numerous functions have been attributed to the cell coat. Although it is not absolutely necessary for the integrity of the cell and for the permeability of the plasma membrane, it nevertheless performs functions that are of great significance to studies in cell biology.

Filtration

The extracellular coats that surround many vertebrate capillaries, especially the kidney glomerulus, act as a filter and regulate the passage of molecules according to size. Hyaluronate in connective tissue may control diffusion.

Microenvironment

The glycocalyx may change concentration of different substances at the cell surface by acting as a diffusion barrier. Because of its charge, the glycocalyx may act as a kind of exchange resin and change the cationic environment of the cell. For example, a muscle cell with its excitable plasma membrane is surrounded by a glycocalyx that can trap sodium ions. Certain components, such as hyaluronate, can drastically change the electrical charge and pH at the cell surface. Because of this, enzymes present at the plasma membrane may change their activity while they are kept in the microenvironment of the cell.

Enzymes

The apical coat is remarkably stable and cannot be separated from the underlying striated border made of microvilli. In the intestinal cells histochemical techniques have demonstrated alkaline phosphatase in the coat as well as on the surface of the microvilli. When these structures are isolated, practically all the enzymes involved in the terminal digestion of carbohydrates and proteins are available in these structure.

Molecular Recognition

The oligosaccharides present in the cell coat may constitute a kind of molecular code for the cell surface. The number of permutations of their individual components (*i.e.*, galactose, hexosamine, mannose, fucose, sialic acid, and so forth) makes for each cell type a special kind of *fingerprint* by which molecular recognition between cells could be established. It is also well known that hormonal receptors and various antigenic properties of the membrane are present in the cell coat.

Blood Group Antigens

The classical division of the ABO blood groups is based on the occurrence of antigens on the red cell surface and of specific antibodies in the serum. The specific ABO antigens are found not only on the erythrocytes but in other tissues and also in secretions. Each one of the groups has a specific structure for the terminal carbohydrates.

Other Antigens

Among these antigens are a those of histocompatibility, which permit the recognition of the cells of one organism and the rejection of other cells that are alien to it (*e.g.*, the rejection of grafts from another organism). This function, essentially, is related to molecular recognition. The major sialoglycoprotein of the red cell membrane carries the M and N antigens that appear, infrequently, in man. It also contains the receptor sites for the influenza virus and various lectins. Molecular recognition reaches its maximum expression in the nervous system, where a neuron can make synaptic contacts with numerous other neurons, thereby forming specific neuronal circuits of immense complexity This property of the cell membrane (*i.e.*, molecular recognition) is dependent on the expression of genes located within the nucleus. Certainly, molecular recognition is one of the fields of cell biology and neurobiology which must be strongly developed in the future.

Cell-cell Recognition

In addition to above cell coat properties are related to the capacity of cells to "recognize" similar cells in a tissue, to adhere to one another, to dissociate and reassociate, and to produce an inhibition in the neighbouring cells by the so-called *contact inhibition*.

Cell Adhesion

There are three main types of contact between cells which lead to their aggregation in a tissue.

1. The aggregation may form by the inclusion of the cell in a common matrix. In most cases this extracellular matrix consists of the coat of the cell. The substances of the coat (cellulose, hyaluronic acid, and so forth) accumulate after the cells have made contact with each other.
2. The aggregation of cells may have little intercellular material. Electron microscopy shows the gap between cells to be frequently on the order of 15 nm. However, even in these small intercellular spaces there is some substance, possibly a mucoprotein, that accounts for the specificity of cell association. Furthermore, the enzymes that are more effective in separating these cells are proteases and mucases.
3. The aggregation may imply the presence of intercellular channels. Cells in a tissue use these channels to interchange more or less freely. In plant cells, cytoplasmic bridges or plasmodesmata have been recognized for a long time. These provide narrow connections between the cells and across the cellulose walls and permit the free passage of ions and probably macromolecules.

Wilson described in 1908 how living sponges forced through a fine silk mesh disaggregate into isolated motile cells, and then, upon standing, reaggregate to form fresh sponges. Much later it was found that embryonic tissues treated with trypsin dissociate into individual cells and then reaggregate to form the specific patterns of the original tissue. The process of reaggregation depends on the motility of cells. If all the cells of a chick embryo are dissociated with trypsin and allowed to stand, they will reaggregate. When similar cells attach to each other, they form aggregates that are characteristic for a given cell population, *e.g.*, retinal cell, kidney, or bone.

The process of reaggregation is not species specific. If cells of chick and mouse embryos are mixed, they reaggregate according to the cell population rather than the species. Reaggregation does not take place in the absence of calcium ions or if the treatment used to disaggregate the tissue selectively blocks certain carbohydrates and glycoproteins.

A cell can be recognized by the machanism and associate with another of similar kind is a property of the cell surface in which the coat is directly involved. The membrane carbohydrates have been implicated in this mechanism. For example, the morphology and type of growth of a cultured cell may be changed by the addition of certain monosaccharides to the medium. An interesting phenomenon related to cell-cell recognition is the so-called *homing of the lymphocytes,* in which lymphocytes leave the bloodstream and enter the lymphatic tissues. This process is dependent on specific carbohydrates present on the lymphocyte membrane. The course of these cells has been followed by labelling lymphocytes with. Lymphocytes taken from the thoracic ducts of rats and injected into other rats went into the spleen and lymph nodes. Treatment of lymphocytes with neuraminidase or with trypsin caused the lymphocytes to accumulate in the liver. These and other experiments suggest the involvement of glycoproteins in this cell-cell recognition mechanism.

Cancer Cells

Cancer involves uncontrolled cell growth, invasion invasion of other tissues, and *dissemination,* to other sites of the organism by way of *metastasis.* All these characteristics indicate that cancer cells have in some way escaped from the controls that regulate normal growth. Many of the changes observed in such cells can be related to alterations in the plasma membrane and especially, in the cell coat.

Contact Inhibition of Movement

Normal cells growing in tissue culture tend to establish cell contacts by adhesion to neighbouring cells. The morphology at the points of adhesion changes; some kind of electron-dense "plaque" is formed in both contacting cells. At the same time there is a slowing down of the ameboid processes which results in *contact inhibition of movement*. In contrast cancer cells are unable to form adhesive junctions and do not show this type of contact inhibition. Such cells tend to have more ruffles and blebs on the cell surface than normal cells. The differences in the movement of normal and cancerous cells are accompanied by changes in the various microfilaments that bring about cell movement.

Contact Inhibition of Mitosis

Normal cultured cells generally divide every 24 hours, as long as they float freely in the medium. However, when they come in close contact in a monolayer, the rate of mitosis slows down and there is inhibition of cell division. The inhibition depends on some unknown signal between cells in contact, and not on a diffusible substance acting at a distance.

The mitotic rate is not inhidited in cancer cells and in cultures the cells tend to pile up, forming irregular masses several layers deep. *A* shows that normal cultured cells form a monolayer and stop cell division by contact inhibition and that transformed (cancer) cells continue to multiply, giving rise to multilayers. Cancerous cells show less adhesion to a solid support or among themselves and motility is more pronounced. These properties may explain why neoplasms invade other tissues and follow an uncontrolled growth. The loss *of contact inhibition* is easily studies in normal cultured cells that are "transformed" into cancerous cells by oncogenic viruses (*i.e.,* viruses capable of inducing cancer) such as the polyoma or the SV40 (*i.e.,* simian virus 40). The changes occurring in cultured hamster cells after transformation by human adenovirus are remarkable when observed by scanning electron microscopy.

Surface Properties

Changes in the surface properties of cancer cells have been observed. For example, using specific staining procedures for mucopolysaccharides (*i.e.,* Hale's reaction or ruthenium red) a considerable increase of this substance was discovered in cells infected with polyoma virus. These cells also showed an increased electrophoretic mobility. Under the electron microscope it was observed that the gap junctions tend to disappear in cells transformed by oncogenic viruses. This may explain the lack of electrical coupling and contact inhibition between cancer cells mentioned earlier.

Another difference is demonstrated by the use of lectins (cell agglutinating proteins) that bind to glycoprotein receptors at the cell surface. Apparently in cancer cells these receptors tend to diffuse more easily within the lipid bilayer than in normal cells and tend to form patches that agglutinate the cells. This greater diffusibility of receptors, which is also found in normal cells during mitosis, may be due to a disorganization of the microfilaments that attach to the plasma membrane.

Molecular Changes of the Cell Membrane

It had been observed that there is difference in the glycolipids and glycoproteins of normal and cancer cells have been observed. A reduction in gangliosides and in the enzymes of synthesis has been observed. In transformed cells there is a disappearance of certain proteins of the membrane.

Membrane Transport

In virally transformed cells a great increase in sugar uptake has been found. Tumor cells frequently release intracellular enzymes to the medium, indicating leakage from cell membranes. A major protease secreted by cancer cells is called *cell factor* (MW 40,000). This enzyme is able to act on *plasminogen* and activate it to *plasmin*. This is a proteolytic enzyme that dissolves blood clots and also removes exposed protein groups at the cell surface of cancer cells. If plasminogen is removed from the medium the morphology of cancer cells returns to normal. Thus, it is possible that this enzymatic digestion could activate the cancer cell into division.

Membrane Antigens

Frequently tumor cells carry new antigens not present in normal cells. For this reason they can induce an immunologic response that in certain favourable conditions may eliminate the cancer cells.

4

Morphology of Mitochondria

In cell biology, mitochondria is a membrane-enclosed organelle found in most eukaryotic cells. These organelles range from 0.5 to 10 micrometers (μm) in diameter. Mitochondria are sometimes described as "cellular power plants" because they generate most of the cell's supply of adenosine triphosphate (ATP), used as a source of chemical energy. In addition to supplying cellular energy, mitochondria are involved in a range of other processes, such as signaling, cellular differentiation, cell death, as well as the control of the cell cycle and cell growth. Mitochondria have been implicated in several human diseases, including mitochondrial disorders and cardiac dysfunction, and may play a role in the aging process. The word mitochondrion comes from the Greek mitos, thread or chondrion, granule.

Several characteristics make mitochondria unique. The number of mitochondria in a cell varies widely by organism and tissue type. Many cells have only a single mitochondrion, whereas others can contain several thousand mitochondria. The organelle is composed of compartments that carry out specialized functions. These compartments or regions include the outer membrane, the intermembrane space, the inner membrane, and the cristae and matrix. Mitochondrial proteins vary depending on the tissue and the species.

In humans, 615 distinct types of proteins have been identified from cardiac mitochondria, whereas in Murinae (rats), 940 proteins encoded by distinct genes have been reported. The mitochondrial proteome is thought to be dynamically regulated. Although most of a cell's DNA is contained in the cell nucleus, the mitochondrion has its own independent genome. Further, its DNA shows substantial similarity to bacterial genomes.

Mitochondri are characterized by a series of morphologic, biochemical, and functional properties These are present in the cytoplasm of protozoa, plant and animal cells. Among these are their size and shape, visibility in vivo, special staining properties, specific structural organization, lipoprotein composition, and content of a large "battery" of enzymes and co-enzymes that interact to produce cellular energy transformations.

From the physiologic viewpoint, mitochondria are energy transducing systems that recover the energy contained in footstuffs (through the Krebs cycle and the respiratory chain) and convert it -by phosphorylation into the high energy phosphate bond of adenosine triphosphate (ATP). Thus, mitochondria are the "power plants" that produce the energy necessary for many cellular functions.

EXAMINATION OF MITOCHONDRIA

Although the examination of mitochondria in living cells is somewhat difficult because of their low refractive index, they can be observed easily in cells cultured in vitro, particularly under dark-field illumination and phase contrast. This examination has been greatly facilitated by colouration with a dilute solution of *Janus green.* The resultant greenish blue stain is due to me action of the cytochrome oxidase system present in mitochondria, which maintains the vital dye in its oxidized (coloured) form. In the surrounding cytoplasm the dye is reduced to a colourless leukobase.

Fluorescent dyes, which are more sensitive, have been used in isolated mitochondria and recently in intact cultured cells. Such dyes are suitable for metabolic studies of mitochondria in situ. This has been demoustrated by the micromanipulation that mitochondria is relatively stable, and without alternation it can be displaced by microneedles. Micromanipulation has demonstrated that mitochondria are relatively stable and can be displaced by the microneedle without alterations. The *specific gravity* is greater than that of the cytoplasm. By ultracentrifugation of living cells at 200,000 to 400,000 g, mitochondria are deposited intact at the centrifugal pole.

Observation in vivo is of particular interest when supplemented by time-lapse cinematography. In cultured fibroblasts, continuous and sometimes rhythmic changes in volume, shape and distribution of mitochondria can be observed.

The two main types of mitochondrial motion are *displacement* from one part of the cell to another, mainly by cytoplasmic streaming, and *bending* and *stretching cycles*. It has been found that the zones of bending are constant and correspond to regions in which the mitochondrial crests are vesicular. Active changes in volume and shape of mitochondria may be caused by chemical, osmotic and mechanochemical changes. In living cells, low amplitude contraction cycles associated with oxidative phosphorylation have been observed.

Shape of Mitochondria

The *shape* of mitochondria is variable, but in general these organelles are *filamentous or granular*. During certain functional stages, other derived forms may be seen For example, a long mitochondrion may swell at one end to assume the form of a *club* or be hollowed out to take the form of a *tennis racket*. At other times mitochondria may become *vesicular* by the appearance of a central clear zone. The morphology of mitochondria varies from one cell to another, but it is more or less constant in cells of a similar type or in those performing the same function.

Size of Mitochondria

The size of mitochondria is also variable; however, in most cells the width is relatively constant (about 0.5 mm), and the length is variable, reaching a maximum of 7 *p.m.* Depending on the functional stage of the cell, however, it possible to find rods that are either thinner or thicker. The size and shape of the fixed mitochondria depend also on the osmotic pressure and the pH of the fixative. Using buffered fixatives at physiologic pH has been shown to be important.

Distribution of Mitochondria

Mitochondria are, in general, uniformly distributed throughout the cytoplasm, but there are many exceptions to this rule. In some cases, they accumulate preferentially around the nucleus or in the peripheral cytoplasm. Overloading with inclusions, such as glycogen and fat, displaces these organelles. During mitosis, mitochondria are concentrated near the spindle, and upon division of the cell they are distributed in approximately equal number between the daughter cells.

The distribution of mitochondria within the cytoplasm should be considered in relation to their function as energy suppliers. In some cells they can move freely, carrying ATP where needed, but in others they are located permanently near the region of the cell where presumably more energy is needed. For example, in certain muscle cells (*e.g.*, diaphragm), mitochondria are grouped like rings or braces around the I-band of the myofibril.

In the rod and cone cells of the retina all mitochondria are located in a portion of the inner segment. The basal mitochondria of the kidney tubule are intimately related to the infoldings of the plasma membrane in this region of the cell. It is assumed that this close relationship with the membrane is related to the supply of energy for the active transport of water and solutes.

Orientation of Mitochondria

Mitochondria may have a more or less definite orientation. For example, in cylindrical cells they are generally oriented in the basal-apical direction, parallel to the main axis. In leukocytes, mitochondria are arranged radially with respect to the centrioles.

It has been suggested that these orientations depend upon the direction of the diffusion currents within cells and are related to the organization of the cytoplasmic matrix.

Number of Mitochondria

Mitochondria are found in the cytoplasm of all eukaryotic cells but not in prokaryotes. In bacteria the respiratory enzymes are located in the plasma membrane. The mitochondrial content of a cell is difficult to determine, but, in general, it varies with the cell type and functional stage. It is estimated that in liver, mitochondria constitute 30 to 35 per cent of the total protein content of the cell, and in kidney, 20 per cent. In yeast, ultrastructural evidence, based on serial sections, points to the existence of a single branched organelle per cell.

A normal liver cell contains between 1000 and 1600 mitochondria, but this number diminishes during regeneration and also in cancerous tissue. This last observation may be related to decreased oxidation that accompanies the increase to anaerobic glycolysis in cancer. Another interesting finding is that there is an increase in the number of mitochondria in the muscle after repeated administration of the thyroid hormone, thyroxin. An increased number of mitochondria has also been found in human hyperthyroidism. D oöcytes contain as many as 300,000 mitochondria—the largest number recorded for a cell. There are fewer mitochondria in green plant cells than in animal cells, because some of their functions are taken over by chloroplasts.

STRUCTURE OF MITOCHONDRIA

A mitochondrion consists of two membranes and two compartments, the larger of which contains the *mitochondrial matrix*. An outer limiting membrane, about 6 nm thick, surrounds the mitochondrion. Within this membrane, and separated from it by a space of about 6 to 8 nm, is an inner membrane that projects into the mitochondrial cavity complex infoldings called *mitochondrial crests.*

This inner membrane, also about 6 nm thick, divides the mitochondrion into two chambers or compartments:

1. The outer chamber contained between the two membranes and in the core of the crests
2. The inner chamber, bound by the inner membrane.

This inner chamber is filled with a relatively dense proteinaceous material usually called the *mitochondrial matrix*. This is generally homogeneous, but in some cases it may contain a finely filamentous material, or small, highly dense granules. These granules are now considered to be sites for binding divalent cations, particularly Mg^{2+} and Ca^{2+}. The mitochondrial crests that project from the inner membrane are, in general, incomplete septa or ridges that do not interrupt the continuity of the inner chamber; thus, the matrix is continuous within the mitochondrion.

The Mitochondrial Matrix Contains Ribosomes and a Circular DNA

Within the mitochondrial matrix are e-mail ribosomes and a circular DNA. The outer and inner membranes and the crests can be considered to be fluid molecular films with a compact molecular structure, the matrix is gel-like and contains a high concentration of soluble proteins and smaller molecules. This double (solid-liquid) structure is important in providing an explanation for some of the mechanical properties of mitochondria.

F_1 Particles

The use of negative staining has enabled recognition of other details of mitochondrial structure. If a mitochondrion is allowed to swell and break in a hypotonic solution and is then immersed in phosphotungstate, the inner membrane in the crest appears covered by particles of 8.5 nm that have a stem, linking each with the membrane.

These so-called "elementary" or "F_1" particles, are regularly spaced at intervals of 10 nm on the inner surface of these membranes.

According to some estimates, there are between 10^4 and 10^5 elementary particles per mitochondrion. These particles correspond to a special ATPase involved in the coupling of oxidation arid phosphorylation.

The presence of such F_1 particles on the matrix side (M side) confers to the inner mitochondrial membrane a characteristic asymmetry which is of basic importance to its function.

It is interesting to recall here that bacterial plasma membranes contain F_1 particles with a similar asymmetric distribution (*i.e.*, towards the bacterial matrix). On the other hand, the opposite orientation is found in chloroplasts where there are F_1 particles on the outer or stromal side of the thylakoid membrane.

Different Cell and Mitochondrial Sturctural Variation

Some investigators assume that a common pattern of mitochondrial structure developed at an early stage of evolution and was subsequently transmitted, without considerable modifications, from protozoa to mammals and from algae to flowering plants. Detailed structural variations can be observed, however. The crests may be arranged longitudinally or they may be simple or branched, forming complex networks. In protozoa, insects, and adrenal cells of the glomerular zone, the infoldings may be tubular instead of lamellar and the tubules may be packed in a regular way.

The number of crests per unit volume of a mitochondrion is also variable. Mitochondria in liver and germinal cells have few crests and an abundant matrix, whereas those in certain muscle cells have numerous crests and little matrix. In some cases the crests are so numerous that they may have a quasi-crystalline disposition. The greatest concentration of crests is found in the flight muscle of insects. In general, there seems to be a correlation between the number of crests and the oxidative activity of the mitochondrion.

On majore feature of the mitochondria is the enlargement of the space within the crests or tubules which, corresponds to the outer compartment of the mitochondrion. This is very conspicuous and appears as tubular or vesicular openings, which are much less opaque than the inner matrix. These structures seem to be related to the specific activity of the gland. In this case, mitochondria, in addition to functioning in cell oxidations, are actively engaged, in the synthesis of steroid hormones.

MITOCHONDRIA AND THE LIPID

Various authors since the time of Altmann have observed that the *disposition of lipids* may be related to mitochondrial activity. In pancreas and liver cells, for instance, after a short period of starvation, the mitochondria come into contact with lipid droplets. The relationship may be so tight that only the inner mitochondrial membrane can be seen adjacent to the lipid in some regions. The electron microscopic images suggest that an active process of fat utilization takes place under the action of the fatty acid oxidases present in mitochondria.

Intramitochondrial Inclusions

An accumulation of pigment derived from hemoglobin has been observed in mitochondria of amphibians. Ferritin molecules accumulate within mitochondria in subjects suffering from Cooley's hereditary anemia.

Another example is the transformation of mitochondria into yolk bodies in eggs of the mollusks. In these, the masses of protein molecules may assume a regular crystalline disposition.

Mitochondrial Sensitivity to Cell Injury and Degeneration

In situation of injury to the cell mitochondria is provide sensitive indicator. Although mitochondria may be readily altered by the action of various agents, the changes are, within certain limits, reversible. If the alteration reaches a certain critical point, however, it becomes irreversible, and this is generally considered degeneration of the mitochondria.

Essentially there are three types of change:

1. Fragmentation into granules, followed by lysis and dispersion,
2. Intense swelling with transformation into large vacuoles,
3. A great accumulation of materials with transformation of mitochondria into hyalin granules. This last change is characteristic of the so-called cloudy swelling and hyalin degeneration that frequently results in cellular death.

A relatively frequent observation in an otherwise normal cell is the presence of degenerating mitochondria in foci of autolysis, constituting a type of lysosome (cytolysosome). Another type of degeneration is the fusion of mitochondria to form large bodies called *chondriospheres.* This degeneration has been found in patients with scurvy and seems to be normal in the adrenal gland of the hamster.

ISOLATION OF MITOCHONDRIAL MEMBRANES

Morphologically homogenous fractions of mitochondria may be isolated from liver, skeletal muscle, heart, and other tissues. Since the early investigations of Hogeboom and others it has been demonstrated that the mitochondrion has a lipoprotein compositon—65 to 70 per cent protein, and 25 to 30 per cent lipid. Most of the lipid content consists of phosphatides (*e.g.*, lecithin and cephalin); cholesterol and other lipids are present in small amounts. Ribonucleic acid was consistently found in about 0.5 per cent of the dry weight.

In recent years the two mitochondrial membranes and the compartments they limit have been separated by density gradient centrifugation. The outer membrane can be separated by causing a swelling, with breakage followed by a contraction of the inner membrane and matrix. The outer membrane is much lighter and needs stronger centrifugal forces or a less dense gradient to be separated.

Mitoplast has been produced by separating these two elements from the outer membrane with digitonin. The mitoplast has pseudopodic processes and is able to carry out oxidative phosphorylation. This separation has provided a clearer definition of the exact localization of the mitochondrial enzyme systems and has revealed interesting differences between the two membranes.

Structural and Chemical Changes in Outer and Inner Membranes

The outer membrane fraction has a 40 per cent lipid content (compared to 20 per cent in the inner membrane), contains more cholesterol, and is higher in phosphatidyl inositol; on the other hand, it is lower in cardiolipin. A fundamental difference between the outer and inner membrane is in the lipid/protein ratio (*i.e.*, about 0.8 in the outer membrane and about 0.3 in the inner membrane). This low lipid/protein ratio indicates that in the inner membrane there is a greater degree of intercalation of the protein within the lipid bilayer. This fact of structure has been confirmed by electron microscopic observations with freeze-fracturing methods. From the morphologic viewpoint the outer membrane lacks the elementary particles that, are prominent in the inner membrane.

Mitochondrial Enzymes are Highly Compartmentalized

The *outer membrane of mitochondria* contains an NADH-cytochrome-c-reductase system which consists of a flavoprotein and cytochrome b_5. The most specific enzyme system of the outer membrane is *monoamine oxidase,* which may serve as an enzyme marker. This membrane also contains kynurenine hydroxylase, fatty acid co-enzyme A ligase, a phospholipase, and various enzymes of the phospholipid metabolism. The supernatant, obtained after treatment with digitonin is considered to originate from the outer compartment between the two membranes. In this compartment are found adenylate kinase, nucleoside diphospho-kinase, DNAase I, and 5'-endonuclease.

The *mitoplast fraction* (inner membrane plus matrix) contains the components of the respiratory chain and oxidative phosphorylation and the soluble enzymes present in the matrix. The *inner membrane* carries all the components of the respiratory chain and the oxidative phosphorylation system (see later discussion). It also has glycerol-phosphate dehydrogenase, choline-dehydrogenase, and several carriers *or translocators* for the permeation of phosphate, glutamate, aspartate, ADP, and ATP. It is calculated that the respiratory chain represents about 20 per cent and the phosphorylating system about 15 per cent of the total protein content of the membrane. These impressive figures give an idea of the importance of these systems in the inner membrane and mitochondrial crests.

The inner membrane contains a remarkably high concentration of cardiolipin (polyglycerophosphatides), which appear to be important in all systems involving electron transport. The *inner compartment* or *mitochondrial matrix* contains all the soluble enzymes of the citric acid or Krebs cycle and those involved in the oxidation of fatty acids. The matrix also contains DNA, ribosomes, and other RNA species and enzymes involved in the synthesis of protein (see later discussion).

The Inner Membrane

It has been proved by various studied that the inner mitochondrial membrane has two major portions, one corresponding to the *mitochondrial crests* and the other to the region that lines the *outer membrane.* For example, it has been reported that the F_1, particles and the coupling factors are present only in the cristae. Furthermore, the activities of several enzymes (*i.e.*, succinate dehydrogenase, b-hydroxybutyrate dehydrogenase, rotenonesensitive NADH-cytochrome-c-reductase, and ATPase) predominate in the cristae.

Studies in recent years have focused attention on the fact that mitochondrial membranes have a net electronegative charge due to anionic sites (*i.e.*, the membranes move towards the anode in an electrophoretic field). Using *polycationic ferritin* as an electron microscopic probe for studying the density and distribution of anionic ionic sites, it was found that while the crests have few sites, there are patches of high density in the regions lining the outer membrane. These areas may correspond to those found in the localization of anionic glycoproteins.

Another regional difference of the inner membrane involves its association with the outer membrane at the *contact zones.* These are observed especially when the mitochondria are in the condensed state; there are about 100 such contacts per liver mitochondrion. At each of these points is a concentration of cytoplasmic ribosomes, which has led to the suggestion that the contact zones could be the sites through which cytoplasmic polypeptides enter the mitochondrion.

FUNCTION OF MITOCHONDRIA

The various functions of mitochondria are so intimately related to their molecular structure that one topic cannot be studied without the other. This section requires some knowledge of the enzymatic mechanisms in which mitochondria are involved. To know the complexity of the enzyme system in numerical terms, in a mitochondrion more than 70 enzymes and co-enzymes and numerous cofactors and metals essential to mitochondrial functions work together in an orderly fashion. Besides oxygen, the only fuel that a mitochondrion needs is phosphate and adenosine diphosphate (ADP); the principal final products are ATP plus CO_2 and H_2O.

The three major foodstuffs of the cell (carbohydrate, fat, and protein) are ultimately degraded in the cytoplasm to acetate, a two-carbon unit that is bound to co-enzyme A to form acetyl co-enzyme A. This penetrates the mitochondrion, and the acetate group condenses with *oxaloacetic acid* to form citric acid, a six-carbon compound. Citric acid is oxidized and loses two carbons as CO_2. In this way the four-carbon compound *succinic acid* is formed, and later on it is oxidized to oxaloacetic acid, starting a new cycle. At each turn of the circle one molecule of acetate penetrates and two CO_2 are released (for more details on the Krebs cycle see textbooks of biochemistry).

An important consequence of the Krebs cycle is that at each turn four pairs of hydrogen atoms are removed from the substrate intermediates by enzymatic dehydrogenation. These hydrogen atoms (or the equivalent pairs of electrons) enter the respiratory chain, being accepted by either NAD^+ or FAD. Three pairs of hydrogens are accepted by NAD^+, reducing it to NADH, and one pair by FAD, reducing it to $FADH_2$ (this last pair comes directly from the succinic dehydrogenase reaction).

Since it takes two turns of the cycle to metabolize the two acetate molecules that come by glycolysis from one molecule of glucose, a total of six molecules of NADH and two of $FADH_2$ are formed at the starting points of the respiratory chain.

Electrons Flow Along a Cytochrrome Chain

The early work of Warburg and Keilin using spectrophotometric techniques led to the concept that cell oxidations are brought about by electron carriers arranged in a chain of increasing *oxidation-reduction potential*. Keilin employed a simple hand spectroscope to study insect muscle and found that there were special pigments which underwent changes with oxidation and reduction during respiration. Such pigments, which had special absorption bands in the reduced state, were called *cytochromes* and were later on demonstrated to be *iron-containing proteins*.

In each cytochrome the iron may be in either a reduced (Fe^{2+}) or an oxidized state (Fe^{3+}), and the electrons flow along a cytochrome chain in which the redox potentials become more positive. With the exception of *cytochrome c*, which is a peripheral protein that can be extracted with salt solutions, these pigments are tightly bound to the inner mitochondrial membrane (as intrinsic proteins; and detergents are needed to solubilize and extract them.

The electron transport chain does not start with the cytochromes but with the co-enzyme *nicotine adenine dinucleotide* (NAD^+), which is reduced to NADH. The couple NAD^+/NADH has the most negative redox potential (Eo = -0.32 volts) of the chain and the highest free energy (*G* = 52.7 kcal/mol), which is released upon oxidation. The reversible NAD^+/NADH system is followed in the chain by a co-enzyme Q_{10} (ubiquinone), a flavoprotein containing FMN and FAD, and then by the series of cytochromes to end with O_2, which has the most positive redox potential (Eo = 0.82) with no free energy to be utilized (G = 0).

This sequence of electron transfer reactions is consistent with the redox potentials of the various electron carriers, which become more positive as electrons pass from the substrate to oxygen. Furthermore, sensitive spectrophotometric methods developed by Chance have confirmed the functional nature of this sequence, showing that NAD^+ is the most reduced member of chain and that cytochrome a and a_s (which form part of *cytochrome* oxidase) are the most oxidized.

There are additional components, and electron transport chain is much complex it has been found in recent years. In recent years it has found that the electron transport chain is extremely complex and that there are additional components. Of particular interest is *co-enzyme Q (CoQ) or ubiquinone* (because it is ubiquitous in cells), which is a lipidsoluble benzoquinone with a long side-chain of ten isoprenoid units (thus the name CoQ_{10}) It is thought that CoQ is a kind of shuttling system between the flavoproteins and cytochrome b In addition, the mitochondrial respiratory chain contains enzyme-bound copper and iron-sulfur proteins.

The Respiratory Chain

Previously discussed components of the respiratory chain, as well as those involved in the mechanism of phosphorylation, are integrated within the molecular structure of the inner mitochondrial membrane. It is important to remember that this membrane has a high protein content (60 to 70 per cent) and that most of these proteins are of the integral or intrinsic type, which are intercalated into the lipid bilayer. In fact it is calculated that one third of this membrane is occupied by phospholipids and the rest by protein.

As in the case of the plasma membrane, there is here a certain degree of *fluidity* that allows for the lateral diffusion of the proteins. This has been demonstrated by freeze-fracturing of the membrane. The study of the distribution of the intramembrane particles under conditions in which there is a slow freezing of the mitochondria shows that this treatment causes aggregation of the particles into patches, while rapid freezing gives a random distribution.

The respiratory chain, together with the phosphorylating system, represents about 35 per cent of the protein of the inner membrane; the rest is represented by the various translocators and structural proteins. By the using mild detergent we can saperate the component of respriratory chain into mulitmolecular complexs. Such complexes have a considerable amount of lipid which is essential for their activity. In fact, if the lipids are extracted, the proteins denature or aggregate, and thus become unable to transfer electrons. Extraction of phospholipids by acetone inactivates some catalytic functions of the membrane; however, restoration of function can be achieved by addition of the lipids and CoQ_{10}

Green and associates have recognized four main complexes (I to IV) which, if mixed in correct stoichiometric ratios, can reconstitute to form the electron transport chain. Complex II corresponds to succinate-Q-reductase (*i.e.*, succinate dehydrogenase), which can transfer electrons directly from succinate to CoQ.

Complex I (NADH-Q-reductase)

NADH is the largest complex, with a molecular weight well above 500,000 and a structure consisting of 15 subunits. It contains as prosthetic groups *Flavin mononucleotide* (FMN) arid six iron-sulfur centres. The NADH reaction site is at the M side; contact between this protein and CoQ *is* apparently made in the middle of the membrane, in the hydrophobic area. Complex I spans the inner mitochondrial membrane and is able to translocate protons across it, from the M side to the C side. This is probably operated through a very hydrophobic subunit of the complex.

Complex II (Succinate-Q-reductase)

Reductase complex is composed of two polypeptides with a total molecular weight of 97,000 daltons. It consists of three iron-sulfur centres and *flavin adenine dinucleotide*. The succinate binding site appears to be on the M side, and the Q site, in the hydrophobic domain of the membrane. In contrast to complex I, succinate-Q-reductase apparently is unable to, translocate protons across the membrane.

Complex III (QH2-cytochrome-c-reductase)

This complex contains a number of subunits with a total molecular weight of about 280,000 daltons. It consists of cytochromes b, cytochrome c_1, and iron-sulfur protein. A proton pump linked to one of the b cytochromes. has been postulated. In the topology of this complex the Q-site may be in the middle of the membrane in the hydrophobic area, and the cytochrome c site, on the C side.

Complex IV (Cytochrome-c-oxidase)

Cytochrome oxidase is a large complex consisting of several polypeptides and having two cytochromes (a and a_3) and two copper atoms. The Molecular weight is about 200,000 daltons. This complex is thought to traverse the mitochondrial membrane, protruding on both surfaces. Such a *transmembranal orientation* is associated with the *vectorial transport of protons* across the membrane. While experimenting with yeast mitochondria it has been shown that cytochrome oxidase is made of seven subunits that can be resolved by gel electrophoresis and that range between 40,000 and 5000 daltons. It is of great biological interest that the three larger polypeptides are very hydrophobic and that their synthesis occurs in mitochondrial ribosomes, and that the four small subunits are relatively hydrophilic and originate in the cytoplasm. The seven subunits are arranged in the membrane in a functional sequence, being in contact with cytochrome c on the C side. The electrons then pass to cytochrome a (also in the C side), then to Cu^{2+} (possibly in the hydrophobic subunits); and finally to cytochrome a_3 and oxygen at the M.

Using antibodies against cytochrome oxidase, it was found that this complex was inhibited when the antibodies were applied to either side of the inner membrane. Such a finding is in agreement with the above-mentioned traverse topography of complex IV. By the use of freeze-fracturing it has been found that cytochrome oxidase, like other integral proteins, is highly mobile in the lateral plane of the membrane. This lateral diffusion is inhibited by the action of cross-linked antibodies.

Electron Transport

The electron transport system of mitochondria is *coupled*

at three points with the phosphorylating system consult. The protons (H^+) originating from electron transfer are translocated by the respiratory chain across the membrane from the M side to the C side.

According to the *chemiosmotic hypothesis,* this translocation creates a pH difference and a membrane potential Both constitute the potential. Both constitute the proton-motive force that tends to move H^+ from the C side back to the M side of the membrane is highly impermeable to H^+ ions, these can only reach the M side through the "proton channel" of the ATPase.

When H^+ moves from the C side to the M side, the F_1-ATPase, operating in reverse, catalyzes ATP synthesis. Conversely, when the F_1-ATPase hydrolyzes ATP, it functions as a "proton pump" and ejects H^+ from the M side to the C side.

Mitochondrial ATPase

The mitochondrial ATPase molecule is characterised by a multi-polypeptide complex.

Three parts can be classified:

- The *F_1 particle* or soluble ATPase, which is in the head piece of the matrix side of the mitochondrial crests F_1 contains five associated subunits.
- A complex of very hydrophobic proteins (*i.e.,* proteolipids) localized in the membrane bilayer and probably containing the proton translocating mechanism.
- A protein stalk that connects the two. This portion corresponds to the *oligomycin-conferring protein* (OSCP) and to another protein needed to bind F_1 to the membrane.

Through the elegant experiment of Racker the Function of the phosphorglating system can be best demonstrated. If isolated mitochondria are subjected to ultrasonic fragmentation, vesicles are formed that show a reverse organization ("inside-out" vesicles), with the F_1 particles situated on the outer surface and with he C-side facing the interior of the vesicle. These vesicles are able not only to respire but also to phosphorylate.

After treatment with urea, which removes the F_1 coupling factor, the membranes no longer phosphorylate. Reconstitution of the complete phosphorylating system can be achieved by mixing the non-phosphorylating membranes with the F_1 particles. After it is isolated, the F_1 particle causes the hydrolysis of ATP to ADP and phosphate; however, when it is attached to the membrane it functions as a synthesizing enzyme.

Respiratory Chains Topological Organization and the Phosphorylation System

All that was just said about the various complexes of the respiratory chain and the phosphorylating system indicates that the inner mitochondrial membrane is highly asymmetrical in its molecular organization. In other words, there is a special sidedness" of the most important components. Under the electron microscope the sideness has been demonstrated by the use of antibodies and also by reagent a that produce electron-dense deposits in the mitochondrial crests.

By using both a special tetrazole, which is converted to formazine by succinate dehydrogenase and 3′,3′-diaminobenzidine (DAB), which is oxidized by cytochrome c, it has been shown that the dense products accumulate on the outside surface of the mitochondrial crest.

From experimental evidence it can be demonstrated that the electron transport system is accessible to NADH and succinate only from the inner matrix side, while, cytochrome c is reached from the outer surface of the membrane. In addition to this transverse topology there is also a lateral topology of the components of the respiratory chain and oxidative phosphorylation. The current theory is that the compact assemblies of macromolecules are arranged in a mosaic which is regularly spaced of by other proteins and lipids. The number of assemblies varies according to the tissue and the folding of the membrane. A mitochondrion from liver has about 15,000 assemblies, while one from the flight muscle of an insect may have as many as 100,000 Within this assembly or *unit of phosphorylation* the main enzymes are in equimolecular quantities. This finding has resulted from the use of ultrarapid spectrophotometric studies and from the determination of binding sites of drugs that specifically affect the various components of the system. For example, *aurovertin* binds to the ATPase; *oligomycin* acts at the level of the binding of ATPase to the membrane; *antimycin* is an inhibitor acting on the QH_2 cytochrome-c-reductase; *rotenone* acts on NADH-Q-reductase; and *cyanide* has its well known poisoning affect on cytochrome-c-oxidase.

The minimum unit of oxidative phosphorylation contains numerous proteins with a total molecular weight approaching 5106 daltons. This unit or assembly must, therefore, cover a considerable area of the membrane. Form biochemical and electron microscopic studies it is possible to imagine a mosaic organization in which each unit os phosphorylation should cover a surface area of about 20 × 20 nm.

The Chemiosmotic Hypothsis

The above-mentioned studies have demonstrated that for each pair of electrons received from the Krebs cycle and transported by the electron transfer system there is one molecule of ATP synthesized by the ATPase. The fundamental problem is that of establishing the nature of the link between these two systems.

The main feature of the *chemiosmotic coupling* proposed by mitchell is that the link is essentially electrochemical in nature. This model is based on the fact that the mitochondrial membrane is impermeable to H^+ and OH^- ions, so that pH differences, acting as energy-rich gradients, can be produced across, it. In this hypothesis the electron transport system is organized in "redox loops" which translocate protons from the M side onto the C side of the membrane. This vectorial movement of protons gives rise to a pH difference and an electrical potential across the membrane.

The chemiosmotic hypothesis postulates that the primary transformations produced by the respiratory chain guide the *osmotic work* needed to accumulate ions. The energy generated by electron transport is conserved in the energy-rich from of a H^+-ion gradient. The pH difference due to the high concentration of OH^- at the M side and of H^+ at the C side also maintain the electrical gradient.

The action of the proton pump of the ATPase, drives the oxidative phosphorylation of ADP to from APT, by which mechanism free energy is conserved:

$$ADP + Pi = ATP + H_2O$$

In this reaction H_2O is also formed because of the dehydration which leads to the removal of H^+ and OH^- ions.

A special feature of this hypothesis is that both the respiratory chain and the ATPase may operate independently. Since coupling by electrical fields can operate at *long-range* distansces there is no need for direct contact between the macromolecules involved. In recent years much experimental evidence has supported the validity of Mitchell's chemosmotic hypothesis, and the mechanism of the redox proton pump has been elucidated by experiments in which various components of the electron transport and phosphorylating system were reconstituted into vesicular bilayers, the so-called liposomes.

The Chemical

An alternate theory—the chemical-conformational hypothesis-postulates that the electron transport and phosphorylating systems are in molecular contact and that information is transmitted by short-range interactions which may, in part, be electrostatic in nature. It is postulated that upon acceptance of the two electrons, each electron transfer protein undergoes a conformational change which is then transmitted to the ATPase by the short-range interactions.

Mitochondria and its Permeability

Metabolic activities of mitochondria take place within the inner mitochondrial compartment, there should be a rapid and active flow of certain metabolites across the two membranes. The products of extramitochondrial metabolism, for example, must reach the mitochondrial matrix in order to undergo oxidation and ADP and phosphate must enter to from ATP. Simultaneously, end products such as H_2O, ATP, urea, and ammonia must leave the mitochondrion.

There are important differences in permeability between the mitochondrial membranes. The outer membrane is freely permeable to electrolytes, water, sucrose, and molecules as large as 10,000 daltons. The inner membrane, is normally impermeable to ions, sucrose.

ADP, ATP, and Pi- Transport by Specific Carriers

The inner mitochondrial membrane uses specific *carriers* or *permeases* for the translocation of various substances. .

In mitochondria the existence of specific carriers for ATP (or ADP), phosphate, succinate (or malate), isocitrate, glutamate, aspartate, and bicarbonate has been suggested. A carrier mechanism has been postulated for Ca^{2+} Mn^{2+},or Sr^{2+}.

From above given the dominant carriers are those involved in the *transport of ADP, ATP, and Pi* (inorganic phosphate) through the inner mitochondrial membrane. Such mechanisms provide for the entrance of ADP and Pi into the mitochondrial matrix and for the exit of ATP to the cytoplasm. Study of the ADP-ATP translocation is facilitated by the fact that there are specific inhibitors of this transport, the experimental use of which has pointed to the *gated pore* as the most plausible of the models.

Such an ADP-ATP carrier would be asymmetrically disposed through the membrane, with two conformations, one facing the C side, and the M side. The same carrier could be used for the reverse translocation of ADP and ADP. Isolation of the carrier protein has been achieved by the use of ^{35}S *carboxyatractylate* (^{35}S CAT), which inhibits transport at the C side, followed by the use of detergents. In this way, a ^{35}S CAT-carrier complex having a molecular from mitochondria. Two of these subunits probably constitute the carrier.

Mitochondrial Swelling and Contraction

During swelling the mitochondrial volume may increase three to five times its normal value in the absence of ADP.

Mitochondria again reach to its origenal size with the addition of ATP or restoration. Studies on the phenomenon of swelling demonstrate that mitochondria have an important function in the uptake and extrusion of intracellular fluid.

Agents known to induce swelling are phosphate, Ca^{2+}, reduced glutathione, and, in particular, the thyroid hormone thyroxine. Thyroxine is the most effective swelling agent and is capable of producing swelling at physiologic concentrations.

Mitochondrial Conformation

Mitochondria may alter their internal conformation between the two extreme states. One is the so-called *orthodox* state that is usually observed in intact tissues. The other corresponds to the *condensed* state, in which there is a dramatic contraction of the inner compartment of the mitochondria accompanied by accumulation of fluid in the outer compartment. In the orthodox conformation the inner membrane shows the characteristic crests; the matrix fills practically the entire volume of the mitochondrion and has a reticular or granular aspect. In the condensed conformation the inner membrane is folded and the matrix, now more represents only about 5 per cent of the mitochondrial volume. In this state, at certain points, it is possible to see contact zones between the inner and outer membranes of the mitochondrion.

The electron transport system is required for the change from condensed to orthodox conformation to take place. Inhibition of the respiratory chain by cyanide, antimycin A, or Amytal will impair this transformation. The orthodox state is induced when the external ADP becomes low and there is none left to be phosphorylated. If at this time DAP is added, respiration is rapidly enhanced, and the contraction of the inner membranes takes place.

It is thought that during the transition from the orthodox to the condensed stage there is change both in the inner membrane and in the matrix. The inner membrane is believed to contain the contractile elements or "mechano-enzymes," but the possible role of the matrix in this contraction should also be considered.

Mitochondrial Accumulation of Ca^{2+} and Phosphate

Respiration and oxidative phosphorylation are the most important functions of mitochondria, another related function is the accumulation of cations. Ca^{2+} can by concentrated in isolated mitochondria up to several hundred times normal values. Phosphate enters the mitochondria along with Ca^{2+}. The amounts of Ca^{2+} and phosphate accumulated may be so great that the dry weight may increase by 25 per cent, and microcrystalline, electron-dense deposits may become visibles within the mitochondria. Electron microscopy coupled with high resolution micro-incineration have been used to study these deposits. This process usually occurs in the osteoplasties present in tissues undergoing calcification.

In the presence of Ca^{2}, mitochondria no longer phosphorylate but infact accumulate Ca^{2+} and phosphate. Both oxidative phosphorylation and accumulation of Ca^{2+} depend on the maintenance of cell respiration (*i.e.,* the Krebs cycle). The accumulation of Ca^{2+} and other cations inside the mitochondrion is accompanied by loss of H^+, so as to maintain an electrical equilibrium. The matrix becomes more alkaline because of OH^- accumulation, while acid is simultaneously released to the outside. Interestingly enough, chloroplasts behave precisely in the reverse direction by pumping out OH^- and accumulating H^+.

Mitochondria and its Biogenesis

Two main mechanisms for the biogenesis of mitochondria have been postulated: origination by division from parent mitochondria or origination *de novo from* simpler building blocks. Mitochondria are distributed between the daughter cells during mitosis, and their number increases during interphase. It has been observed with time-lapse cinematography that mitochondria gradually elongate and then fragment into smaller mitochondria. This observation has been verified in *Neuropora.*

After labeling of a choline deficient mutant of *Neurospora* with radioactive choline, the radioactivity was followed in the mitochondria of the second and third generations. By radioautography it is found that all mitochondria of the original progeny were labeled. The mitochondria of each daughter cell were also labeled but contained about half the radioactivity. This seemed to indicate that mitochondria had divided and grown by the addition of new lecithin molecules to the existing mitochondrial framework. Yeast cells grown anaerobically lack a complete respiratory chain (cytochromes b and a are absent); under the electron microscope they show no typical mitochondria; When the yeast cells are placed in air, however, the mitochondrial membranes present fuse, unfold, and from true mitochondria that contain the cytochromes. This observation is somewhat in agreement with the *de novo* synthesis theory of the biogenesis of mitochondria.

The chemical composition of the outer membrane is similar to that of the endoplasmic reticulum and very different from that of the inner membrane Furthermore, continuities between the outer membrane of the mitochondrion and the endoplasmic reticulum, and even with the axon membrane, have also been observed.

Mitochondria are Semiautonomous Organelles

Since mitochondria and chloroplast biogenesis contain DNA as well as ribosomes and are able to synthesize the protein that way studies in this matter has been enhance during recent years. The term *semiautonomous organelles* was applied to the two structures in recognition of these findings. However, at the same time, the name indicated that the biogenesis was highly dependent on the nuclear genome and the biosynthetic activity of the ground cytoplasm.

It is now known that the mitochondrial mass grows by the integrated activity of both genetic systems, which cooperate in time and space to synthesize the main componets. The mitochondrial, DNA codes for the mitochondrial, ribosomal, and transfer RNA and for a few proteins of the inner membrane. However, most of the proteins of the mitochondrion result from the activity of the nuclear genes and are synthesized on ribosome of the cytosol. The study of mitochondrial biogenesis involves knowledge of many concepts that are considered in later chapers dealing with gene expression. Fore this reason at this point we can treat this subject in only a very elementary way.

Mitochondrial DNA

Chevremont had observed that under certain conditions, mitochondria of cultured cells give a positive Feulgen reaction, until 1963 the general opinion was that DNA is exclusively confined to the nucleus. In that year M. and S. Nass observed filaments within mitochondria that they interpreted as DNA molecules. This finding was fully confirmed in cell sections, as well as in DNA extracted and studied by the surface spreading technique. The mitochondrial DNA is localized in the inner compartment or matrix and is probably attached to the inner membrane at the point where DNA duplication starts. This duplication is under nuclear control and the enzymes used are imported from the cytosol. A single mitochodrion may contain one or more DNA molecules, depending on its size; *i.e.,* the larger the mitochondrion, the more DNA molecules present. Thus, a normal eukaryotic cell has at least as many copies mitochondria. DNA as it has mitochondria. It has been calculated that in an adult man there may be 10^{17} mitochondria molecules, an immense number; it would be interesting to know whether they are similar or different in their nucleotide sequence. So far the results obtained with various techniques suggest that they are identical and that genes are represented in single copies.

In most animals and plants the mitochondria DNA is circular, although it may be highly twisted. In most animal cells mitochondria DNA is only 5.5 μm long, while in yeast it may be longer (*i.e.,* 26 μm). Because of its higher guanine-cytosine content, mtDNA has a higher temperature than nuclear DNA. It is relatively easy to separate the strands of mitochondria DNA and to localize genes after bybridization with RNA molecules.

Mitochondria DNA from human cells is a circular DNA molecule of about 15,000 base pairs. A significant proportion of this DNA is devoted to coding for the protein synthesizing machinery. MtDNA codes for the mitochondrial ribosomal RNAs (12S and 16S), for about 19tRNAs (probably a full set is encoded, *i.e.*, tRNAs for all the amino acids), and for the mRNAs of about 20 proteins. Most of the proteins for the mitochondrial ribosomes come from the outside and are encoded by the nuclear genes. The mRNAs transcribed from the mitochondrial DNA code mostly for highly hydrophobic proteins. It seems that the reason for having all this biosynthetic machinery in side the mitochondrion is probably that there is no other way of getting these very hydrophobic proteins inside the organelle.

Mitochondrial Ribosomes

In yeast and *Neurospora, ribosomes* have been assigned to a 70S class similar to that of bacteria; in mammalian cells, however, mitochondrial ribosomes are definitely smaller and have a total sedimentation coefficient of 55S, with subunits of 35 and 25S. Ribosomes in mitochondria appear to be tightly associated with the membrane. In spite of their low sedimentation coefficient (55S), animal ribosomes are slightly larger than the 70S bacterial; this is because the animal structures contain more protein.

Mitochondria and Chloroplasts are Intracellular Prokaryotic Parasites

At the end of the last century, cytologists like Altmann and Schimper speculated on purely morphological grounds that mitochondria and chloroplasts might be intracellular parasites that had established a *symbiotic* relationship with the eukaryotic cell. Bacteria were thought to have originated the mitochondria, and blue-green algology, with the recognition that these organelles have a certain degree of autonomy, this symbiont hypothesis has been reframed.

According to the revised of as an anaerobic organism deriving its energy from glycolysis, a process that occurs in the cytoplasmic matrix and the parasite contains reactants of the Krebs cycle and the respiratory chain and is able to carry on respiration and oxidative phosphorylation. The symhiont hypothesis is even more plausible in the case of plant cells, since the parasite would be the chloroplast, an autotrophic microorganism able to transform energy from light.

The symbiont hypothesis is based on the many similarities between mitochondria, chloroplasts, and prokaryotes, which when considered from an prokaryotes, which when considered from an evolutionary viewpoint, may be more than circumstantial. There are similarities in the *localization of the respiratory chain*. In bacteria this is situated in the plasma membrane, which also has the F_1-ATase projecting into the matrix. Certain bacteria have membranous projections extending from the plasma membrane forming the so-called *mesosomes. Such* membranous projections, which are comparable to mitochondrial crests, have been separated and shown to contain the respiratory chain. The inner mitochondrial membrane and the matrix may represent the original symbiont enclosed within a membrane of cellular origin.

The *mitochondrial DNA* is circular, as it frequently is in chromosomes of prokaryons. Mitochondria contain ribosomes that are smaller than those belonging to the cell-at-large; however, they are generally smaller than those in bacteria. As just mentioned, protein synthesis in mitochondria and in bacteria is inhibited by chloramphenicol, whereas the extramitochondrial protein synthesis of the higher cell is not affected.

In mitochondria we have also already noted that there is evidence of a DNA-dependent RNA synthesis, which indicates partial autonomy of this organelle. The amount of information carried by the mitochondrial DNA, however, is insufficient for an autonomous biogenesis. In support of the possible prokaryotic origin of mitochondria and chloroplast is the fact that *intracellular symbiosis* may contain certain in nature. Thus, *paramecia* may contain certain bacteria, and blue-green algae may occur in simple animals. The endosymbiosis of a photosynthetic prokaryote confers upon the host the ability to capture light energy to synthesize various products. On the other hand this provides for the prokaryote a constant environment in which to grow and reproduce. In evolutionary terms it is possible to conceive that a symbiotic relationship could have evolved into the present situation, in which the organelles have only a certain degree of autonomy and depend on the nucleus and cytosol of the cell for the synthesis of most of their specific components.

5

Mechanical Function of Cell

Some cell protuberances has been correlated to the orientation and distribution of microtubules. They are considered to be a framework which determines the shaping of the cell and redistributes its content. The integrity of microtubules is necessary for the individual shape of many cells and the rigidity of elongated processes. These particularly evident in axons and dendrites.

MORPHOGENESIS

Cells's mechanical function is the role that microtubules play in the shaping of the cell during *cell differentiation*. For example, the-elongation of the cells during the induction of the lens placode in the eye is accompanied by the appearance of numerous microtubules. The morphogenetic changes that occur during spermiogenesis provide another interesting example. The enormous elongation that takes place in the nucleus of the spermatid is accompanied by the production of an orderly array of microtubules that are wrapped around the nucleus in a direction perpendicular to the nuclear axis and that have a double helical arrangement.

CELLULAR POLARITY AND MOTILITY

The determination of the intrinsic polarity of certain cells is also related to the mechanical function. Treatment of various culture cells with Colcemid results in a change in motion.

The following forms of movement have been observed to persist membrane:

- Ruffling

- Endocytosis,
- Attachment to the surface,
- Extension of microvilli.

Only a saltatory movement of particles was found to be inhibited after the destruction of the microtubules. However, the directional gliding of the cell is replaced by a random movement.

Regarding culture cells they are treated in vitro with dibutyryl cyclic AMP become elongated and the number of microtubules increases. All these observations suggests that microtubules are responsible for maintenance of the shape of the cell, the polarity of movement, and the distribution of certain or ganelles.

CIRCULATION AND TRANSPORT

May also be involved in the *transport* of macromolecules in the cell's interior; to this end they probably form channels in the cytoplasm. A very interesting example is the protozoan *Actinosphaerium* (Heliozoa) which sends out long thin pseudopodia within which cytoplasmic particles migrate back and forth. These pseudopodia contains as many as 500 microtubules disposed in a helical configuration. When these protozoans are exposed to cold, or high pressure the pseudopodia are withdrawn and the microtubules depolymerize.

Another example of an association between microbules and transport of particulate material is the melanocyte, in which the melanin granules move centrifugally and centripetally with different stimuli. The granules have been observed moving between channels created by the microtubules in the cytoplasmic matrix. The erythrophores found in fish scales the pigment granules may move at a speed of 25 to 30 μm per second between the microtubules.

Sensory Transduction

Regularly arranged bundles of microtubules are common in sensory receptors, and a possible function in transduction of different incident energy has been attributed to them.

MICROTUBULAR ORGANELLES

Many cells organs can be considered to be derived from special assemblies of microtubules. In fact, cilia, flagella, basal bodies, and centrioles have, as one of their main components, groups of microtubules arranged in a special fashion. For this reason and in spite of the complexity of their structure, these organelles will be considered at this point, along with some of their physiological aspects.

Cillary and Flagellar Motions

Cillary and flagellar all motility is adapted to liquid media and is executed by minute, specially differentiated appendices that vary in size and number. They are called *flagella* if they are few and long, and *cilia,* if short and numerous.

In Protozoa, especially the Infusoria, each cell has hundreds or thousands of minute cilia, and their movement permits a rapid procession of the organism in the liquid medium.

In some special regions of the Infusoria, several cilia fuse and form larger conical appendices, *the cirri,* or membranes observed as *undulating membranes.*

Out of entire class of Protozoa, the Flagellata, is characterized by the presence of flagella. The spermatozoa of metazoans move as isolated cells by means of flagella. On the other hand, epithelial cells that possess vibratile cilia and constitute true ciliated sheets are relatively common. These may cover large areas of the external surface of the body and determine the motion of the animal. Such is the case with some Platyhelminthes and Nemertea and also with larvae of Echinodermata, Mollusca, and Annelida.

More often, the ciliated epithelial sheets line cavities or internal tubes, such as the air passages of the respiratory system or various parts of the genital tract. In these organs all the cilia move simultaneously in the same direction, and fluid currents are thus produced. In some cases, the currents serve to eliminate solid particles in suspension (*e.g.,* in the respiratory system). The eggs of amphibians and mammals are driven along the oviduct with the aid of vibratile cilia.

The Cillary Apparatus

Some of the essential components of cillary apparatus are:

- The *cilium,* which is the slender cylindroid process, limited by the plasmalemma (cell membrane), that projects from the free surface of the cell,
- The *basal body,* or granule, the intracellular organelle similar to the centriole from which it originates,
- In some cells fine fibrils—called *ciliary rootlets*—that arise from the basal granule and converge into a conical bundle, the pointed extremity of which ends at one side of the nucleus.

Regarding the basal bodies they are embedded in the ectoplasmic layer beneath the cell surface. In general, they are spaced uniformly and in parallel rows. Cilia and flagella are extremely delicate filaments whose thickness is often at the limit of the resolving power of the light microscope.

Various epithelia have appendices similar in shape to cilia, but immobile; these are called *stereocilia.* Examples are the processes of the epithelial cells of the epididymis. In the macula and crista of the inner ear, there are stereocilia in addition to motile cilia, or *kynocilia.* Stereocilia do not contain microtubules.

The Axoneme

Regarding the term Axoneme the some is applied to the axial basic microtubular structure of cilia and flagella, and is it thought to be the essential motile element. In various flagella *accessory structures* made of fibres, microtubules, crystalline bodies, or mitochondrial derivatives may be found. The axoneme of a cilium or flagellum may range from a few microns to 1 to 2 millimeters in length, but the outside diameter is about 0.21 μm. It is generally surrounded by an *outer ciliary membrane,* which is continuous with the plasma membrane. All the components of the axoneme are embedded within the *ciliary matrix.*

With the fundamental 9 + 2 microtubular pattern and the other essential structural components of the axoneme. A plane perpendicular to the line joining the two central tubules divides the axoneme into a right and a left symmetrical half. It is generally admitted that the plane of the ciliary beat is perpendicular to this plane of symmetry.

The pairs of peripheral microtubules have an ellipsoidal profile, whereas the central ones are circular. The diameter of the peripheral tubules varies from 18 to 25 nm with the central tubules being slightly larger. The two microtubules of each peripheral pair can be clearly distinguished by several morphologic features. The doublets are skewed at about 10 degrees so that one tubule, designated subfibre A, lies closer to the axis than the other (*i.e.*, subfibre B). The microtubule of subfibre A is smaller, but complete, whereas that of subfibre B is larger and incomlete, since it lacks the wall adjacent to A. In fact, while A has 13 tubulin subunits, B has only 11. Furthermore, subfibre.

A has processes—the so-called *dyein arms* —that are oriented in the same direction in all microtubules. This orientation is clockwise when the axoneme is viewed from base to tip.

Dynein Arms

The arms of subfibre A which are generally called *dynein arms* because they contain dynein, a higher molecular weight ATPase (300,000 to 400,000 daltons). This is a Mg^{2+}-and Ca^{2+} -activated enzyme, which after solubilization can recombine at the same position on the A microtubules. Two isoenzymatic forms of the ATPase, dynein I and II, have been separated by electrophoresis.[16] Dynein I constitutes the largest proportion of the axonemal ATPase and is present in the arms of subfibre A. As will be mentioned later, the interaction between tubulin and dynein is thought to underlie the basic mechanism of ciliary and flagellar motion.

Nexin Links

The doublets are linked by *interdoublet* or *nexin links*. This name is derived from the fact that another protein called *nexin* has been isolated from these links. This protein has a molecular weight of 150,000 to 160,000 daltons. The function of these links is unknown, but they may serve as structures that maintain the geometric integrity of the axoneme during the sliding motion.

Radial Spokes

The spokes terminate in a dense knob or head, which may have a forklike structure. Certain observations that the spokes are attached perpendicularly to the ciliary axis where it is straight and that they are relatively detached in bent or tilted regions of the axis has led to the hypothesis that they may be active in the conversion of active sliding between the outer doublets into local axial being.

Basal Bodies (Kinetosomes) and Centrioles

After the classic works of Hennegay and Lenhosesk in the year 1897, it has been suggested that basal bodies (or kinetosomes) of cilia and flagella are homologous with the centrioles found in mitotic spindles. In some cells it was discovered that a centriole engaged in mitosis could carry a cilium at the same time. This homology was fully confirmed with the introduction of electron microscopy. Centrioles are cylinders that measure, on the average, 0.2 μm–0.5 μm; at times they may be as long as 2 μm. This cylinder is open on both ends, unless it carries a cilium. In the later case, it is separated from the cilium by a *ciliary plate.*

In freeze-fractured cilia it has been possible to observed the so-called ciliary necklace. This consists of two to six rows of membrane particles present just below the ciliary plate where the central pair of microtubules end.

In the wall of the centriole there are nine groups of microtubules arranged in a circle. Each group is a triplet formed of three tubules (rather than two, as in cilia) that are skewed towards the centre. Since the angle made by the set of tubules with the tangent to the cylinder varies from 70 degrees in the proximal end to 20 to 30 degrees in the distal end, the tubules twist from one end to the other or describe a helical course. The tubules are designated *subfibre A, B,* and *C.* from the centre towards the periphery.

Both subfibres A and B cross the ciliary plate and are continuous with the corresponding subfibres in the axonemene; subfibre C terminates near the *ciliary* or *basal plate.* There are no central microtubules in the centrioles and no central microtubules in the centrioles and no special arms; however, they are linked by connectives. The proximal portion of the centriole has a cartwheel appearance with spokes that radiate from a central hub and connect to subfibre of each triplet. The presence of the cartwheel structure at the proximal end provides the centriole with a structural and functional polarity. The growth of the centriole is from the distal end, and in the case of kinetosomes, it is from this end that the cilium is formed. Further more, the *procentrioles,* which are formed at right angles to the centriole, are located near the proximal end. Both the above given centrioles may give rise to a cilium but in most cases only one does, and the sister centriole remains at a right angle to the kinetosome. In most multiciliated cells there are single basal bodies. It will be shown later that in sensory cells having nonmotile ciliary derivatives and a 9 + 0 configuration, there are two basal bodies. The two centrioles of a pair are always located at a distance from each other of at least 0.8 μm and are perpendicular to each other.

Ciliary Rootlets

In some of the cells, ciliary rootlets originate from the basal body. Most rootlets are striated, having a regular cross-banding with a repeating period of 55 to 70 nm. Five intraperiodic sub-bands have been observed in the cross-striated rootlets. The striated fibres are composed of parallel microfilaments, 3 to 7 nm in diameter, which in turn are formed of globular subunits. These fibres and filaments may serve a structural role such as anchoring the kinetosomes. By analogy with other microfilaments, a contractile role has been postulated for the ciliary rootlets. Furthermore, ATPase has been found associated with the cross-bands of certain rootlets.

Basal Feet and Satellites

Basal feels are dense processes that are arranged

perpendicularly to the basal body in a particular direction and originate from two to three of the triplets. These processes impose a structural asymmetry on the basal body that has been related to the direction of the ciliary beat. The basal foot, which is composed of microfilaments that terminate in a dense bar, may be a focal point for the convergence of microtubules. *Satellites* or *pericentriolar bodies,* are electron-dense structures lying near the centriole that probably are nucleating sites for microtubules.

Ciliary Movement

Ciliary movement can be analysed easily by scraping the pharyngeal epithelium of a frog or toad with a spatula and placing the scrapings in a drop of physiologic salt solution between a slide and a coverglass. On the free surface of the epithelial cell, the rapid motion of the vibratile cilia can be seen. If a row of cilia is observed, the contraction is *metachronic* in the plane of the direction of motion; that is, it starts before or after the contraction of the next cilium. In this way true waves of contraction are formed.

Now consider the other part that, in a plane perpendicular to the direction of motion, the contraction is *isochronic;* all the cilia are observed in the same phase of contraction at a given time. The rhythmic contraction of cilia has been interpreted in different ways. A two-step process that involves *intraciliary* excitation followed by *interciliary* conduction has been proposed to explain the metachronic rhythm of the ciliary beat. This mechanism evidently does not depend on the nervous system, since it persists after the epithelium has been separated from the rest of the organism. However, cytoplasmic continuity is indispensable to its maintenance, for if a cut is made in the row of cilia the waves of contraction of the two isolated pieces becomes uncoordinated.

If a piece of epithelium is removed from the pharynx of a frog and implanted with a reversed orientation, the movement is maintained but in the opposite direction. Ciliary contractions are generally rapid (10 to 17 per second in the pharynx of the frog). Analysis of the motion has been facilitated greatly by stroboscopic and ultrarapid microcinematography. Ciliary movement may be pendulous, unciform (hooklike), infundibuliform, or undulant. The first two are carried out in a single plane. In the pendulous movement, typical of the ciliated Protozoa, the cilium is rigid and the motion is carried out by a flexion at its base.

On the other hand, in the unciform movement, the most common type in the Metazoa, the cilium is doubled upon contraction and takes the shape of a hook. In the infundibuliform movement, the cilium or flagellum rotates, describing a conical or funnel-shaped figure. In the undulant motion, characteristic of flagella contraction waves proceed from the site of implantation and pass to the free border. It was shown above that the plane of the beat of a cilium is perpendicular to that passing through the central microtubules.

Although the effective stroke is a single plane, the cilium moves away from the original plane during the recovery phase. This movement, which can be beautifully demonstrated by scanning electron microscopy,[23] implies that either the whole cilium rotates at its base or that the central microtubules rotate.

Since the last century, ciliary and flagellar motion have been compared to that of muscle. However, the complex structure of the axoneme suggests that this motion may be more complex than the planar sliding of myofilaments in muscle.

The key process appears to be the sliding of the microtubular doublets over each other, with the associated making and breaking of the cross-bridges between adjacent doublets, represented by the dynein arms. The sliding of doublets during the bending of the cilium or flagellum has actually been observed and measured. This displacement of the doublets has been determined because the central microtubules extend to the very tip, whereas subfibres B terminate at a distance from the tip of 1 to 2 μm, and subfibres. A terminate slightly beyond subfibres B. This arrangement has also made it possible to demonstrate that the two subfibres do not move relative to one another within the doublet. In the diagram of the amount of sliding displacement is shown to depend on the distance separating the tubules and on the bend angle.

Experiments on ciliary motion has been aimed at demonstrating the involvement of ATP; it was mentioned above that the ATPase activity is associated with the protein *dynein* in the arms of subfibres. A Dynein constitutes about 8 per cent of the total ciliary protein. In glycerin-extracted cilia, flagella, and spermatozoan tails, the addition of ATP produces rhythmic activity that may persist for a few minutes or even hours. Interaction between tubulin and dynein seems to be essential in this type of cell motility. When studying myofilaments in muscle, we shall see that the actin-myosin system is the other process involved in contraction.

Several approaches are current by being used to study the molecular mechanisms in ciliary motion. One of them has been the use of sperm flagella in which the membranes have been removed by detergents. These sperm can be reactivated with ATP. Extraction of dynein arms by KC1 leads to a reduction in the frequency of beating, which can be restored to the original level by the addition of *dynein*. An anti-dynein antiserum inhibits the movement of spermatozoa without membranes, and ATP induces, it. The other observation is that, unlike muscle, cilia and flagella do not require Ca^{2+} for activation. In many cases, however, Ca^{2+} has a regulating effect on the wave form of the beating.

Kartagener's Syndrome

One of the recent approach to the study of ciliary motion is the use of mutants of *Chlamydomonas* in which various deficiencies in axoneme structure have been observed. From the medical point of view, this research is of great importance in the so-called *Kartangener's syndrome,* in which there is infertility with nonmotile sperm chronic bronchitis, and sinusitis. It was been demonstrated that this syndrome is due to a mutation involving the dynein arms. In these patients *cilia and sperm are non-motile because they lack dynein arms, which contain ATPase.*

Photoreceptors

Regarding the studies on the structure of retinal rods and cones have shown that the short, fibrous connection found between the outer and inner segments is of a ciliary nature. Cross sections of this so-called connecting cilium have revealed nine pairs of filaments similar to those found in cilia. In this structure, however, there are no central microtubules.

The outer segment of this cilium is composed of numerous double membrane rod sacs arranged like a stack of coins. The first stage in the development of a rod is a primitive cilium projecting from the bulge of cytoplasm that constitutes the primordium of the inner segment. This cilium contains the nine pairs of filaments and the two basal centrioles.

The apical end is filled with a vesicular material. In the second stage the apical region of the primitive cilium enlarges greatly, owing to the rapid building up of the vesicles and cisternae that constitute the primitive rod sacs. The proximal part of the primitive cilium remains undifferentiated and constitutes the connecting cilium of the adult.

Another structures have also been recognized as ciliary derivatives. For example, the so-called crown cells of the saccus vasculosus found in the third ventricle of fishes are modified cilia with swollen ends that are filled with vesicles. Also, the primitive sensory cells of the pineal eye found in certain lizards have a ciliary structure.

Cilia and Flagella Originate from Basal Bodies

The origin of centrioles and basal bodies, or kinetosomes, is viewed from the standpoint that these are possibly semiautonomous cell organelles and in this way similar to mitochondria and chloroplasts. Isolation of kinetosomes from *Tetrahymena* has verified the presence of RNA and DNA, and this suggests that they are capable of some protein synthesis. More direct evidence of DNA in centrioles was obtained by use of the fluorescent dye acridine orange and by ^{3}H-thymidine incorporation followed by treatment with DNAse. However, most of these studies refer to the pellicle, and the exact localization of DNA has not been determined.

Furthermore, these findings do not resolve the question of whether the synthesis of ciliary proteins depends on the DNA-RNA components of the centriole.

Some of the early cytologists realised that the development of cilia and flagella was directly related to the presence of centrioles. Flagella may be experimentally removed and their regeneration followed; thus, in *Chlamydomonas* it was found that the growth rate of a flagellum is about 0.2 μm per minute. During the studies of the origin of centrioles and kinetosomes were hampered by the small size of these structures. Light microscopists thought that centrioles originate by division from preexisting centrioles and that they possessed a genetic continuity comparable to that of chromosomes. The electron microscope has failed to confirm that such a division process takes place.

We find several in directions that centrioles may be generated by two different mechanisms. Centrioles destined to form mitotic spindles of single cilia arise directly from the wall of the preexisting centriole. The *daughter centrioles* appear first as annular structures (*procentrioles*), which lengthen into cylinders. There are three groups of tubules originate from single and double groups that first appear at the base of the procentriole. When they are half grown the daughter centrioles are released into the cytoplasm to complete their maturation. Usually, although not always, daughter centrioles are formed one at a time. Another mechanism in the formation of centrioles is found in those destined to become kinetosomes, as in a ciliated epithelium. The centrioles are assembled progressively from a precursor fibrogranular material located in the apical cytoplasm. The newly formed centrioles become aligned in rows beneath the apical plasma membrane, and each centriole may then produce satellites from the side, a root from its base and a cilium from its apex.

Cilium's development begins with the appearance of a vesicle that becomes attached to the distal end of the centriole. The growing ciliary shaft invaginates the vesicular wall, which forms a temporary ciliary sheath until the permanent one is formed. In many epithelial cells short cilia enclosed within vesicles may be observed. Under the electron microscopic study of centrioles have revealed that they have a specific tubular structure, that they do not divide, and that they are found in association with other tubular structures. This relationship suggests that the function of the centriole may be to regulate the synthesis and organization of microtubules within the cytoplasm.

Microfilaments

We have studied earlier that the ground cytoplasm or cytosol contains a kind of *cytoskeleton* formed by microtubules and various types of microfilaments. From what was said about the function of cytoplasmic microtubules the reader may have gathered the impression that they play mainly a mechanical function, constituting a kind of *passive portion or framework of the cytoskeleton*. This concept may not be completely accurate, since these microtubules could store energy by distortion, and during disassembly they could release elastic forces that might transport cytoplasmic structures. Furthermore, we have seen before that the interaction between the tubulin of microtubules and dynein is the main mechanism of motion of cilia and flagella.

Pleasantly the *active* or *motile function of the cytoskeleton* has been attributed to some of the microfilaments that are observed by electron microscopy as thin filaments between 6 and 10 nm thick. Various studies have suggested that such filaments represent contractile systems which may be involved in cytoplasmic streaming, ameboid motion, and other types of cell activities in which a motive force is involved. These studies have recently led to very important conclusions with the demonstration that the contractile proteins of muscle (*i.e., actin* and *myosin*) are also widely found in nonmuscular systems and that they may be chemical components of some of the microfilaments.

Microtrabecular Lattice in Cytosol

High voltage electron microscopy has helped in identifying microfilaments in whole culture cells and in obtaining a three-dimensional view of the so-called *microtrabecular lattice.*

Just beneath the membrane there are bundles of filaments that can be seen to be made of actin and that are in continuity with a lattice of similar filaments that pervades the cytosol. The channels or space found in between this *trabecular lattice* are of the order of 50 to 100 nm and in the living cell may provide for the rapid diffusion of fluid and metabolites through all parts of the cytosol.

The filaments of the lattice are in contact with vesicles of the endoplasmic reticulum, with microtubules and with polysomes, all of which seem to be contained of supported within the lattice. On the other hand, mitochondria appear to be free of the lattice, and this may explain the fact that they move rather freely in the cytosol. This image of the cytoskeleton explains the solgel transformation that occurs in the cytosol. The polymerization and depolymerization of this lattice may also provide for small movements of the contained organelles.

The same observations have been made on thin sections of various cells. Networks of microfilaments have been observed in the growth cones of neuroblasts and in the undulating membranes of cultured cells. A thin sheath of microfilaments forming a network is observed below the plasma membrane of fibroblast and glial cells. In nerve endings a three dimensional network of actin filaments has been observed.

Cytochalasin

During the study of microtubules the depolymerizing effect of colchicine has been of paramount importance, in the case of microfilaments the action of the drug *cytochalasin B* has had a decisive influence. Although the mechanism of action is still unknown, cytochalasin B has been found to impair numerous cell activities in which some types of microfilaments are involved. For example, it inhibits smooth muscle contraction, beat of heart cells, migration of cells, cytokinesis, endocytosis, exocytosis, and other processes.

During application to the alga Nilella, cytochalasin B produces rapid cessation of cyclosis, which is reinitiated after removal of the drug. Cytochalasin B disrupts the regular arrangement of microfilaments associated with some of these functions. For example, the contractile ring of microfilaments observed during cell cleavage is altered by this drug. Not all microfilaments, however, are sensitive to the drug. These and other studies have led to the general conclusion that the cytochalasin-B-sensitive microfilaments are the contractile machinery of non-muscle cells.

Actin, Myosin, and other Contractile Protein

The contractile proteins *actin* and *myosin* are present in non-muscle cells has for the first time which is suggested a molecular mechanism for cellular movements. It is now known that all eukaryotic cells contain high concentrations of *actin* and low concentrations of *myosin,* the two *force-generating proteins.* In addition, there are other proteins, such as *tropomyosin* and *troponin* that, as in muscle, play a regulatory function. Finally, there are *actin-binding* proteins that can cross-link action filaments, and a-*actinin,* a protein found in the Z-line of muscle and which may participate in actin-filament-membrane interactions.

Actin

It comprises a large proportion of the cytoplasmic proteins of many cells; in developing nerve cells, it may constitute up to 20 per cent of these proteins. Actin is present principally in its globular form (G-actin) with a molecular weight of 45,000 and it may quickly polymerize to form the microfilaments of fibrous actin (F-actin). This G to F-actin transition seems to represent the basis of the classical sol-gel transition in the cytoplasm of moving cells.

Cytroplasmic action is very similar in structure to muscle actin and forms identical 6-nm wide microfilaments consisting of a double helical array of globular actin molecules. Actin filaments can be identified under the electron microscope by virtue of the fact that they become *decorated* with myosin (heavy meromyosin) to form arrow-like complexes. *Myosin* has been found in cells having ameboid motion, such as amebae and slime molds, and also in platelets.

Myosin

If we compare with actin, myosin is found in a low concentration and is rather heterogenous in size and composition. In most cases it has a molecular weight of 460,000 daltons, similar to that of muscle myosin, but the two molecules contain different subunits. In *Acanthamoeba* there is a *"minimyosin"* with a MW of 180,000 daltons. In spite of these differences, all myosins bind reversibly to actin filaments and contain a Ca^{2+} activated ATPase. As in muscle, both these functions are localized in the globular heads of the molecule.

Identification of myosin filaments with the electron microscope, at present, is less satisfactory. However, filaments that are 14 to 22 runs thick and 0.7 μm long, which are of probable myosin nature, have been observed.

Contractile and Regulatory Proteins

The microfilaments can be detected by fluorescent antibodies against *actin, myosin, tropomyosin,* and *a-actinin.* In culture cells actin has a typical pattern of straight and often parallel fibres that are predominantly localized on the lower, contacting side of the cell and that are in close contact with the plasma membrane. This pattern is completely disorganized by the action of cytochalasin B, which does not affect the microtubules.

Staining of culture cells anti-myosin antibodies produces a somewhat similar image to that of actin; however, the majority of the filaments show interruptions or striations. Staining with anti-a-actinin antibodies also reveals a similar pattern, with a periodic arrangement along the filament. These results suggest that *a*-actinin (and also *tropomyosin*) may be involved in the organization of the actin bundles of microfilaments and in the attachment of them to the plasma membrane, especially in the areas in which there is contact with a solid substrate or with other cells. In these contacting regions actin microfilaments are arranged in parallel, forming bundles that in a living spreading cell may be visible as *stress fibres*. This is very interesting that the microvillus of the intestine, which contains a bundle of actin filaments that is attached to the plasma membrane by way of a dense material that stains with antibodies against a-actinin. This actin bundle has a polarization similar to that found in amuscle sarcomere except that in this case the Z-line is replaced by the a-actinin at the tip of the microvillus.

Two Recognizable Types of Microfilaments

In addition to microtubules (20 to 30 nm) and actin microfilaments (5 to 7 nm), in a variety of cells there is a third cytoskeletal components composed of filaments of an intermediate size (7 to 12 nm). These have been identified in fibroblasts, muscle cells, and glial cells. Other examples include the *neurofilaments* of nerves, the *tonofilaments* of epithelia, and the *prekeratin filaments* of cells in which keratinization occurs.

Studies with fluorescent antibodies in culture cells have shown the presence of a system of wavy fibres, some of which lie near the cell boundary. This system of fibres is affected neither by cytochalasin, which disorganizes actin filaments, nor by Colcemid, which depolymerizes microtubles. This *intermediate system* of microfilaments appears to be related to the anchorage and maintenance of the epithelioid cell shape in nerve axons the neurofilaments could be involved in axoplasmic transport. Isolation of intermediate filaments is facilitated in cell cultures of baby hamster kidney treated with colchicine. In these cells the 10-nm filaments aggregate to form juxtanuclear caps, which can be isolated. Two proteins of 54,000 and 55,000 daltons were resolved by SDS gel electrophoresis of these intermediary filaments.

The basis of contraction in non-muscular systems is still hypothetical, but the scheme that is evolving is in some ways similar to that of muscle. In non-muscular systems it is based on the interaction of actin and myosin filaments, with the consequent production of a shearing force. The differences between the mechanism here and in muscle are in the more random distribution of these proteins and the much smaller concentration of myosin. These factors may help to account for the much slower concentration in these primitive contractile systems. The attachment of active filaments to the plasma membrane, how- ever, is favourable to the production of shearing forces acting at the level of the membrane.

Many motile events in non-muscle cells, Like cell spreading and shape changes, locomotion, membrane ruffling, cytokinesis, phagocytosis, and formation of pseudopodia, require a tight coupling between the cell surface and the contractile apparatus, and this could be achieved by a-actinin molecules that bind the actin microfilaments to the plasma membrane

Against muscle cells, in which actin filaments are stable, most non-muscle cells have transitory filaments that may appear or disappear within one minute. Thus the cell must have some mechanism that controls the assembly and disassembly of the actin monomers and the formation of microfilament bundles. Such mechanisms are altered in cells undergoing transformation by oncogenic viruses some of which may show a reduced number of bundles.

We shall now consider two classic types of cell motility in which micro filaments and actinmyosin interactions are involved. This is generally found in the plant cell and another called *ameboid motion* which is found in animal cells and in certain protozoa.

Cytoplasmic Streaming (Cyclosis)

Cytoplasmic streaming, or *cyclosis,* is easily observed in plant cells, in which the cytoplasm is generally reduced to a layer next to the cellulose wall and to fine trabeculae crossing the large central vacuole. Continuous currents can be seen that displace chloroplasts and other cytoplasmic granules.

In some plant cells the protoplasmic current can be initiated by chemicals *(chemodynesis) or* by light *(photodynesis).* Cyclosis is modified by temperature, by The action of ions, or by change in pH. Cyclosis is stopped by mechanical injuries, electric shock, or some anesthetics. Some auxins (plants growth hormones). Increase the rate of cyclosis. In general, all the factors that decrease cell viscosity increase the speed of protoplasmic current and vice versa. Cyclosis decreases progressively in cells subjected to increased hydrostatic pressure at the same time that protoplasm becomes more liquid.

The classic experimental work on cyclosis has utilized the cylindroid cells of *Nitella,* which have a thin protoplasmic layer of about 15 mm surrounding a central vacuole of 0.5 mm by 10 cm. This protoplasmic layer is divided into a cortical region of structured cytoplasm in which are embedded non-motile chloroplasts and a layer of isolated cytoplasm (the endoplasm) where cyclosis takes place. The whole system is surrounded by the plasma membrane, under which is a single layer of microtubules.

Bundles of actin filaments of about 0.2 mm are situated just beneath the chloroplasts in the cortical region and in a direction that is parallel to that of cyclosis. (Each bundle is composed of about 100 microfilaments of 5 to 6 nm.) It is assumed that the separation of the ectoplasm and endoplasm is labile and the actin filaments may be incorporated into the endoplasm.

It has been proved that plant cytoplasm also contains a myosin-like protein. Thus, the actual motive force may be provided by a mechanism of actin-myosin interaction that takes place in the region between the stationary ectoplasm and the moving endoplasm. Furthermore, it has been found that cytoplasmic streaming is stopped when the cortex of the *Nitella* is treated with cytochalasin B.

Ameboid Motion

In ameboid motion the cell changes shape actively, sending forth cytoplasmic projections called *pseudopodia,* into which the protoplasm flows. Although this special form of locomotion can be observed easily in amebae, it also occurs in numerous other types of cells. One need only to place a drop of blood between a slide and coverglass to see that the leukocytes, at first spheroidal, change their shape, emit pseudopodia, and move about. In tissue cultures, cells move out actively, forming the zone of migration. These changes also occur in vivo. For example in epithelial repair, the cells free themselves and slide along actively towards the depth of the wound. In an inflammatory process, leukocytes wander out of the blood vessels *(dlapedests)* by active ameboid motion and progress towards the focus of infection.

Some amebae are predominantly *monopodial* but others may be temporarily or permanently *polypodial.* The shape of pseudopodia varies between a stout, almost cylinadrical *lobopodium* and a fine filamentous or branching *filopodium.* Sometimes these fine processes may be anastomosing (resticulopodia), as in Foraminitera.

As shown by the classic studies of Mast, the protoplasm of the ameba has a clear ectoplasm, which expands considerably towards the end of the pseudopodium. The axial endoplasm is surrounded by a "shear zone" where particles move more freely. At the advancing end is the hyaline cap, and just posterior to it, the "fountain zone," where the axial endoplasm appears to contract actively and flows below the ectoplasmic tube. At the opposite end is the tail process, also called the *moid,* and near it, the *recruitment zone,* where the endoplasm is recruited from the walls of the ectoplasm in the posterior third of the cell.

The rate of progression varies between 0.5 and 4.6 mm per second among different amebae. In the neutrophilic leukocytes it is approximately 0.58 mm per second. This rate is modified by temperature and other environmental factors. Insufficient oxygen supply does not stop the movement but does slow it. For this type of locomotion calcium is needed. Serve mechanical injury, electric shock, or ultraviolet radiation causes retraction of pseudopodia.

An important factor in ameboid motion is *adhesion to a solid support.* An ameba that floats freely in the liquid medium can emit pseudopodia but does not progress; only when it adheres to a solid surface does it commence this type of locomotion. There are substances that at influence the motion by attracting or repelling the cells. This property, d is called *chemotaxi.* It has great importance in defence mechanisms. All that has been said thus far about microfilaments and cell motility applies to ameboid motion. Actin filaments and myosin aggregates, together forming thicker filaments, have, been found in amebae, and there seems no doubt that the actin-myosin interaction provides the actual motive force for ameboid motion. However, the views are divided regarding the most likely site of contraction. Whereas some investigators regard the posterior region of the ectoplasmic tube as more active, others give more importance to the "fountain zone" at the advancing end of the ameba.

6

Types of Plant Cell

Plant cells are eukaryotic cells that differ in several key respects from the cells of other eukaryotic organisms. A large central vacuole, a water-filled volume enclosed by a membrane known as the tonoplast maintains the cell's turgor, controls movement of molecules between the cytosol and sap, stores useful material and digests waste proteins and organelles. A cell wall composed of cellulose and hemicellulose, pectin and in many cases lignin, are secreted by the protoplast on the outside of the cell membrane. This contrasts with the cell walls of fungi (which are made of chitin), and of bacteria, which are made of peptidoglycan.

Specialised cell-cell communication pathways known as plasmodesmata, pores in the primary cell wall through which the plasmalemma and endoplasmic reticulum of adjacent cells are continuous. Plastids, notably the chloroplasts which contain chlorophyll and the biochemical systems for light harvesting and photosynthesis, but also amyloplasts specialized for starch storage, elaioplasts specialized for fat storage and chromoplasts specialized for synthesis and storage of pigments.

As in mitochondria, which have a genome encoding 37 genes plastids have their own genomes of about 100-120 unique genes and probably arose as prokaryotic endosymbionts living in the cells of an early eukaryotic ancestor of the land plants and algae. Unlike animal cells, plant cells are stationary. Cell division by construction of a phragmoplast as a template for building a cell plate late in cytokinesis is characteristic of land plants and a few groups of algae, notably the Charophytes and the Order Trentepohliales. The sperm of bryophytes have flagellae similar to those in animals, but higher plants, (including Gymnosperms and flowering plants) lack the flagellae and centrioles that are present in animal cells.

CELL TYPES

Parenchyma cells are living cells that have diverse functions ranging from storage and support to photosynthesis and phloem loading (transfer cells). Apart from the xylem and phloem in its vascular bundles, leaves are mainly composed of parenchyma cells. Some parenchyma cells, as in the epidermis, are specialized for light penetration and focusing or regulation of gas exchange, but others are among the least specialized cells in plant tissue, and may remain totipotent, capable of dividing to produce new populations of undifferentiated cells, throughout their lives.

Parenchyma cells have thin, permeable primary walls enabling the transport of small molecules between them, and their cytoplasm is responsible for a wide range of biochemical functions such as nectar secretion, or the manufacture of secondary products that discourage herbivory. Parenchyma cells which contain many chloroplasts and are concerned primarily with photosynthesis are called chlorenchyma cells. Others, such as the majority of the parenchyma cells in potato tubers and the seed cotyledons of legumes have a storage function.

Collenchyma cells-collenchyma cells are alive at maturity and have only a primary wall and a secondary wall. These cells mature from meristem derivatives that initially resemble parenchyma, but differences quickly become apparent. Plastids do not develop and the secretory apparatus (ER and Golgi) proliferates to secrete additional primary wall. The wall is most commonly thickest at the corners, where three or more cells come in contact and thinnest where only two cells come in contact, though other arrangements of the wall thickening are possible. Pectin and hemicellulose are the dominant constituents of collenchyma cell walls of dicotyledon angiosperms which may contain as little as 20% of cellulose in Petasites. Collenchyma cells are typically quite elongated, and may divide transversely to give a septate appearance. The role of this cell type is to support the plant in axes still growing in length, and to confer flexibility and tensile strength on tissues. The primary wall lacks lignin that would make it tough and rigid, so this cell type provides what could be called plastic support. Support that can hold a young stem or petiole into the air, but in cells that can be stretched as the cells around them elongate. Stretchable support (without elastic snap-back) is a good way to describe what collenchyma does. Parts of the strings in celery are collenchyma.

Sclerenchyma cells-Sclerenchyma cells (from the Greek skleros, hard) are hard and tough cells with a function in mechanical support. They are of two broad types–sclereids or stone cells and fibres. The cells develop an extensive secondary cell wall that is laid down on the inside of the primary cell wall. The secondary wall is impregnated with lignin, making it hard and impermeable to water. Thus, these cells cannot survive for long as they cannot exchange sufficient material to maintain active metabolism. Sclerenchyma cells are typically dead at functional maturity, and the cytoplasm is missing, leaving an empty central cavity.

Functions for sclereid cells (hard cells that give leaves or fruits a gritty texture) include discouraging herbivory, by damaging digestive passages in small insect larval stages, and physical protection (a solid tissue of hard sclereid cells form the pit wall in a peach and many other fruits). Functions of fibres include provision of load-bearing support and tensile strength to the leaves and stems of herbaceous plants. Sclerenchyma fibres are not involved in conduction, either of water and nutrients (as in the xylem) or of carbon compounds (as in the phloem), but it is likely that they may have evolved as modifications of xylem and phloem initials in early land plants.

TISSUE TYPES

The major classes of cells differentiate from undifferentiated meristematic cells (analogous to the stem cells of animals) to form the tissue structures of roots, stems, leaves, flowers and reproductive structures.

Xylem cells are elongated cells with lignified secondary thickening of the cell walls. Xylem cells are specialised for conduction of water, and first appeared in plants during their transition to land in the Silurian period more than 425 million years ago. The possession of xylem defines the vascular plants or Tracheophytes. Xylem tracheids are pointed, elongated xylem cells, the simplest of which have continuous primary cell walls and lignified secondary wall thickenings in the form of rings, hoops or reticulate networks. More complex tracheids with valve-like perforations called bordered pits characterise the gymnosperms.

The ferns and other pteridophytes and the gymnosperms only have xylem tracheids, while the angiosperms also have xylem vessels. Vessel members are hollow xylem cells aligned end-to-end, without end walls that are assembled into long continuous tubes. The bryophytes lack true xylem cells, but their sporophytes have a water conducting tissue known as the hydrome that is composed of elongated cells of simpler construction.

Phloem is a specialised tissue for food conduction in higher plants.The conduction of food is a complex process which is carried in the plant with the help of special cell called phloem cells. These cells conduct inter and intra cellular fluid (food:- proteins and other essential element required by the plant for its metabolism) through the process of osmosis. This phenomena is called as ascent of sap in plants. Phloem consists of two cell types, the sieve tubes and the intimately-associated companion cells.

The sieve tube elements lack nuclei and ribosomes, and their metabolism and functions are regulated by the adjacent nucleate companion cells. Sieve tubes are joined end to end with perforate end-plates between known as sieve plates, which allow transport of photosynthate between the sieve elements. The companion cells, connected to the sieve tubes via plasmodesmata, are responsible for loading the phloem with sugars. The bryophytes lack phloem, but moss sporophytes have a simpler tissue with analogous function known as the leptome. Plant epidermal cells are specialised parenchyma cells covering the external surfaces of leaves, stems and roots. The epidermal cells of aerial organs arise from the superficial layer of cells known as the tunica (L1 and L2 layers) that covers the plant shoot apex, whereas the cortex and vascular tissues arise from innermost layer of the shoot apex known as the corpus (L3 layer). The epidermis of roots originates from the layer of cells immediately beneath the root cap. The epidermis of all aerial organs, but not roots, is covered with a cuticle made of waxes and the polyester cutin.

Several cell types may be present in the epidermis. Notable among these are the stomatal guard cells, glandular and clothing hairs or trichomes, and the root hairs of primary roots. In the shoot epidermis of most plants, only the guard cells have chloroplasts. The epidermal cells of the primary shoot are thought to be the only plant cells with the biochemical capacity to synthesize cutin.

CELL WALL OF PLANT CELLS

There is a kind of *exoskeleton* that provides protection and mechanical support for the plant cell. This includes the maintenance of a balance between the osmotic pressure of the intracellular fluid and the tendency of water to penetrate the cell. When plant cells are placed in a solution that has an osmotic pressure similar to that of the intracellular fluid, the cytoplasm remains adherent to the cellulose wall. When the solution of the medium is more concentrated than that of the cell, it loses water, and the cytoplasm retracts from the rigid cell wall. On the other hand, when the solution of the medium is less concentrated than that of the intracellular fluid, the cell swells and eventually bursts.

The formation, from cambial cells, of *xylem vessels* (which become lignified) or of *phloem tissue* (used to transport to the rest of the plant the material that has been photosynthesized in the leaves)—is mainly the result of the special synthesis and assembly of the cell wall. In some cases, the cell wall, can be compared to a piece of plastic reinforced by glass fibres. The wall consists of a *microfibrillar* network lying in a gel-like *matrix* of interlinked molecules. The microfibrils are mostly *cellulose* (the most abundant biological product on earth) consisting of straight polysaccharide chains made of glucose units linked by 1-4 β- bonds. Macrofibrils visible with the light microscope are composed of microfibrils about 25 nm in diameter, each of which is, in turn, composed of about 2000 cellulose chains. About 100 cellulose chains are held together in an elementary fibril. X-ray diffraction reveals a crystalline pattern in cellulose, with a repeating period of 1.03 nm along the fibber axis. This corresponds to a *cellobiose* unit composed of two *β*-glucose molecules.

Extra than some protein, the matrix contains other major polysaccharide fractions are:

- Pectin substances (containing galactose, arabinose, and galacturonic acid), which are soluble in water,
- Hemicelluloses (composed of glucose, xylose, mannose, and glucuronic acid), which are extracted with alkali.

There is only lignin found in maters cell wall which is made of an insoluble aromatic polymer resulting from the polymerization of phenolic alcohols. Certain cell walls may have *cuticular substances (cutin waxes)* and *mineral* deposits in the form of calcium and magnesium carbonates and silicates. In fungi and yeasts the cell wall is composed of *chitin,* a polymer of glucosamine.

Actually, the cell wall determines the *spape* of cells and serves in the classification of plant tissues. (Consideration of specialized plant tissues is beyond the scope of this book. In general, *Primary* and *secondary* walls develop in a special sequence and are differentiated by the disposition of the cellulose microfibrils and by the composition of the matrix.

Regarding the classification of the layers in a matured cell wall depends upon the knowledge of its development during and after cell division. The *cell plate,* which is formed mainly from vesicles of the Golgi complex that become aligned at the equatorial plane, forms the intercellular layer or *middle lamellae* of the more mature cell wall; this layer contains pectin substances. After cell division each daughter cell deposits other layers that constitute the *primary cell wall,* which is composed of pectin, hemicellulose, and a loose network of the cellulose microfibrils, most of which are oriented transverse to the long axis of the cell.

Incase the cell has enlarged to its matter size, then with the development of large vacuolar, the secondary wall appears. When the cell has enlarged to its mature size, usually with the development of large vacuoles the *secondary wall* appears. This consists of material that is added to the inner surface of the primary wall, either a *homogeneous thickening,* as in cells forming sieve tubes *(phloem),* or as *localized thickenings,* as in xylem vessels. In both cases this material consists mainly of cellulose and hemicellulose, with little pectin.

In xylem development a further differentiation consists of the penetration of *lignin* from the outside into the secondary thickenings. This hydrophobic polymer replaces water and finally encrusts all the microfibrils and the matrix. At this stage the is lignified and the cell dies. Regarding the biogenesis of cellulose and other cell wall components two main pathways have been described. One involves the *Golgi complex,* and the other appears to be directly associated with the *plasma membrane* involvement of the Golgi is very evident in some scale-bearing algae. These scales consist of a radial microfibrillar network associated with a spiral one, and an amorphous wall.

The synthesis of polysaccharides in the Golgi of the algae has been followed by the use of periodic acid and silver methenamine. It is evident that the Golgi membranes with their *glucosyl transferase* content are able to polymerize glucan chains into cellulosic microfibrils. The whole process of synthesis, assembly, and release of the scales at the cell surface has been followed under the electron microscope.

The plasma membrane is the most common site of cellulose synthesis in plant cells. However, one must not disregard the essential functions that are probably carried out by the endoplasmic reticulum and the Golgi complex. It is possible that the glycosyl transferases synthesized in the endoplasmic reticulum could be transferred to the Golgi, and from there to the plasma membrane, where they may become active in the synthesis of microfibrils. Using freeze-fracturing, globular complexes situated at the end of growing microfibrils have been observed in the secondary walls of cotton fibres. These globules have been interpreted as enzyme complexes involved in the synthesis of the microfibrils. Such globules, which are mobile, may be guided by cytoplasmic microtubules.

This is the main characteristic of most plant cell is that the presence of bridges of cytoplasmic material that establish a continuity between adjacent cells. These bridges, called *plasmodesmata,* pass through the thickness of the pectocellulose membrane. They have a thin plasma membrane and many contain tubules that establish continuities between the endoplastic reticulum of both cells. The presence of plasmodesmata permits the free circulation of fluid, which is essential to the maintenance of plant cell tonicity, and probably also allows passage of solutes and even of macromolecules. According to these concepts, cell walls do not represent complete partitions between cells, but constitute a vast syncytium supported by a skeleton formed by the pectocellulose membranes.

Regarding the formation of plasmodesmaty, the same is concerned with the formation of the *cell plate,* mentioned earlier, which appears at the equator of dividing cells during telophase. At this time the cell plate is crossed by vesicles and tubules of the endoplasmic reticulum that determine the location of the plasmodesmata.

CYTOPLASM

In meristematic cells the membranes of the cytoplasmic vacuolar system are relatively scanty and are masked by the numerous ribosomes that fill the cytoplasmic matrix. Indeed, in undifferentiated cells most of these particles are not attaced to the membranes, but are free in the matrix. Microtubules are preferentially localized below the plasma membrane and are oriented tangentially. A role in wall deposition has been postulated for these structures. Microtubules are found in the regions of cytoplasm that underlie the points where the microfibrils of the secondary wall are being deposited. As mentioned earlier, they may play a role in guiding the enzymes involed in the biosynthesis of the microfibrils, and may thus determine the pattern (spiral or reticular) in which the microfibrils are deposited. Reticulum is plant has rough portions with attached ribosomes and smooth sections in animal cells. It has been postulated that the plasma membrane of the plant cell invaginates and actively takes in fluid (pinocytosis). The same process also occurs in numerous animal cells. The great development of the endomembrane system during cell differentiation os related to the intense hydration of the cytoplasm. This process may give rise to huge vacuoles that are filled with liquid and may be confluent. As a result, the cytoplasm may become compressed in a thin layer against the cellulose membrane and may show cytoplasmic movements, called *cyclosis*.

The endoplasmic reticulum plays a role in the biosynthesis of proteins, lipids, and polysaccharides and in the formation of certain organelles, such as *protein bodies, glyoxysomes,* and vacuoles. We mentioned before that the plasmodesmata are crossed by tubules of smooth endoplasmic reticulum. In cells that differentiate to form sieve tubes, pads of *callose,* a polysaccharide composed of glucose units linked by 1-3 β-bonds, is deposited over the endoplasmic reticulum. Thus, this organelle is also related to the formation of sieve pores.

The seed is the resting stage of the plant in which the cells are in a highly dehydrated state. In addition to the primordia of the root and shoot, the seed is composed of food stores in the form of starch, fat, and proteins.

Seed development involves the following three main stages:

1. A *division phase,* in which the cells of the embryo and the cotyledons are formed.
2. A *phase of cell expansion,* in which proteins and other reserves are synthesized and stored in cotyledons.
3. A *drying phase,* in which there is loss of water with further deposition of the reserves.

Several proteins (*i.e., zein* in corn endosperm, *vicilin* and *legumin* in legume cotyledons) ae localized in special organelles called *protein bodies* or *aleurone grains*. During the expansion phase, the cells have an extensive rough endoplasmic reticulum which is involved in the biosynthesis of stored proteins. These are first accumulated in the central vacuole, and later on smaller protein bodies are formed, possibly within cisternae of the endoplasmic reticulum or the Golgi complex. This process has been followed radioautographically by the use of labeled amino acids. *Zein,* the main protein of corn, is synthesized by mRNA that is specifically associated with ribosomes attached to the endoplasmic reticulum. It has been found that protein bodies also have bound ribosomes that may produce zein.

Seed Germination is germination is accompanied by the metabolism of the protein bodies of cotyledons. An *endopeptidase* synthesized in the seedlings penetrates into the protein bodies, causing hydrolysis of the stored proteins. Enzyme synthesis is preceded by proliferation of the rough endoplasmic reticulum. When treated with the hormone *gibberellic acid,* the aleurone cells of cereal endosperm produce α-amylase and other hydrolytic enzymes, which are released extracellulary and attack the starch reserves. The synthesis of these enzymes is correlated with an extensive proliferation of the rough endoplasmic reticulum in these aleurone cells.

Regarding the germination of the castor bean (*ricinus*) is accompanied by breakdown of a large amount of fat stored in the endosperm. This process is accompanied by the development of *glyoxysomes,* special organelles involved in triglyceride metabolism. Glyoxysomes consist of an amorphous protein matrix surrounded by a limiting membrane. Both the phospholipid components and the enzymes of these organelles originate from the endoplasmic reticulum.

The enzyme of the glyoxysome are used to transform the fat stores of the seed the seed into carbohydrates by way of the *glyoxylate cycle* which is a modification of the Krebs cycle. The overall equation of this cycle is as follows 2 acetyl CoA $\rightarrow$ succinate + $2H^+$. The difference between the glyoxylate cycle and the Krebs cycle is that the former used two auxiliary enzymes, *isocitrate lyase* and *malate synthetase,* and requires two molecules of acetyl CoA, instead of one. The other three enzymes of the cycle (*citrate synthetase, aconitase,* and *malate dehydrogenase*) are the same as those present in the Krebs cycle.

All the enzymes of the glyoxylate cycle have been found in isolated glyoxysomes of the castor bean endosperm. When seeds of the castor bean germinate, considerable change is produced involving the synthesis of glyoxysomic enzymes. If the endosperm is fractionated at different states of germination, it is possible to observe that the marker enzyme malate synthetase shifts from fractions corresponding to the endoplasmic reticulum to others representing the glyoxysomes. These findings support the concept that these organelles are made in the endoplasmic reticulum.

At telophase they aggregate at the periphery of the cell plate and form small vesicles, which fuse to form the plate. As in animal cells, the dictyosomes of some plant cells (*e.g.,* root cap cells of maize) are directly related to secretion. The Golgi cisternae become filled with secretion products, which are then concentrated and discharged.

Dictyosomes and their associated vesicles are numerous in cells involved in the synthesis of mucilage, and in those of the outer root cap of the bean.

The mucilage is produced at the expense of the starch bodies present in plastids. The dictyosomes of plant cells contain specific enzymes, such as thiamine pyrophosphatase and inosinic diphosphatase. Incorporation of labeled glucose is highest during the formation of the cell plate. The role of the Golgi apparatus in cell wall formation was described above.

Plant cell's mitochondria have a structure essentially similar to that of those in animal cells. In meristems, mitochondria have relatively few crests and an abundant matrix. During differentiation this internal structure may vary. In cells engaged in photosynthesis (leaf cells), mitochondria show an increased number of crests; in cells containing starch granules (amyloplasts), mitochondria remain undifferentiated, as in meristems.

One important point is the relationship between

mitochondria and chloroplasts. Guillermond studied this problem in leaf meristems and postulated that in early stages there are two types of organelles. One is typically composed of short mitochondria and the other of long filamentous bodies.

These elongated organelles increase in thickness and may give rise to vesicles, starch granules, or chloroplasts, whereas the short mitochondria remain unchanged. According to this view, both types of organelles are independent. The introduction of electron microscopy has, in general, confirmed this viewpoint. Although early plastids, also called proplastids, in some ways resemble mitochondria, they are readily distinguishable because of fewer projections of the inner membrane, their large size, and the presence of dense granules in the matrix.

THE CHLOROPLAST AND OTHER PLASTIDS

Schimper first used the term "plastid" for special cytoplasmic organells present in eukaryotic plant cells in 1883. The most important of all, *the chloroplasts,* are characterized by the presence of pigments such as *chlorophyll* and *cerotenoids* and by their fundamental role in *photosynthesis.*

Extra than chloroplasts, other coloured plastyids may be observed. These are grouped under the name *chromoplasts.* Yellow or orange chromoplasts occur in petals, fruits, and roots of certain higher plants. In general, they have a reduced chlorophyll content and are thus less active photosynthetically. The red colour of ripe tomatoes is the result of chromoplasts that contain the red pigment *lycopene* a member of the carotenoid family. Chromoplasts containing various pigments (*e.g., phyco-erythrin* and *phycocyanin*) are found in algae.

Colourless plastids or leukoplasts are found in embryonic and germ cells. During embryonic development leukoplasts in certain differentiated zones of the root produce starch granules, called *amyloplasts*. They may be seen under the polarization microscope, because of their characteristic birefringence, or they may be distinguished by histochemical reactions for starch. Leukoplasts are also found in meristematic cells and in those regions of the plant not receiving light.

Plastids located in the cotyledon and the primordium of the stem are colourless at first but eventually become filled with chlorophyll and acquire the characteristic green colour of chloroplasts. It has already been studied that the protein bodies, other contain starch, fat (*elaioplasts or oleosomes*), or essential oils. It was noted in *Phaseolus vulgaris,* that a single leukoplast may store both starch and protein; *i.e.,* it may qualify as both an amyloplast and a proteinoplast. The starch granules are deposited in the matrix or stroma of the plastid, while the protein, which sometimes shows a crystalline arrangement, is always accumulated within membrane-bound sacs.

It is the greatest biological importance that chloroplasts are the most common of the plastides. By the process of *photosynthesis,* they produce oxygen and most of the chemical energy used on our planet by living organisms. Life is essentially maintained by the chloroplasts; without them there would be no plants or animals, since the latter feed on the foodstuffs produced by plants. From a human point of view we may say that each molecule of oxygen used in respiration and each carbon atom in our bodies has at one time passed through a chloroplast. Chloroplasts are localized mainly in the cells of the leaves of higher plants and in algae.

In the point of view of evolutionary the first living organism were and appeared at least one billion year before photosynthesis. This process appeared first in prokaryotes and then in eukaryotes. With the release of oxygen, a by-product of photosynthesis, the atmosphere favoured aerobic organisms, and anaerobic organisms became a small fraction of all living forms.

In photosynthetic prokaryotes and in the chloropasts of eukaryotes photosynthesis and carried out in species membrances. Photosynthesis is carried out in special membranes in photosynthetic prokaryotes and in the chloroplasts of eukaryotes. These organelles are able to capture light energy and to transform it into chemical energy in the form of ATP, $NADP^+$, and the various foodstuffs that are produced. There is much difference between the shape of the cells within a species and chloroplasts. But these organelles are relatively constant within cells of the same tissue. In leaves of higher plants, each cell contains a large number of spheroid, ovoid, or discoid chloroplasts. Some are club-shaped, having a thin middle zone and bulging ends filled with chlorophyll. Chloroplasts are frequently vesicular, with a colourless centre. The presence of starch granules is detected by the characteristic blue iodine reaction.

Considerably the size of chloroplastes varies. The average diameter in higher plants is 4 to 6 μm. This is constant for a given cell type, but sexual and genetic differences are found. For instance, chloroplasts in polyploid cells are larger than those in the corresponding diploid cells. In general, chloroplasts of plants grown in the shade are larger and contain more chlorophyll than those of plants grown in sunlight. Algae, such as *chlamydomonas,* often possess a single huge chloroplast. In higher plants there are 20 to 40 chloroplasts per cell. It has been calculated that the leaf of *Ricinus communis* contains about 400,000 chloroplasts per square millimeter of surface area. When the number of chloroplasts is insufficient, it is increased by division; when excessive, it is reduced by degeneration. The *distribution* of the chloroplasts within the cell may vary with the amount of light energy Chloroplasts are sometimes distributed homogeneously within the cytoplasm, but are frequently packed near the nucleus, or close to the cell wall.

Observation of living epidermal cells from leaves of Iris..... And other genera has shown that chloroplasts are displaced and deformed by the action of cytoplasmic streaming (cyclosis). In addition to this passive *motility,* active movements of an ameboid or contractile type, which are sometimes related to the degree of illumination, have been observed.

In chloroplasts isolated from spinach, the changes in shape and volume caused by the presence of eight have been obersed. The volume decreases considerably after the chloroplasts are struck by light and photophosphorylation is initiated; this effect is reversible. In the dark contraction of chloroplasts may be induced by addition of ATP. Two proteins having contractile properties, which may account for this phenomenon, have been extracted from isolated chloroplasts. Chloroplasts apparently multiply by division—elongation of the plastid and constriction of the central portion. They have a higher density than the cytoplasm and migrate to the centrifugal pole of the cell when submitted to the action of centrifugal force.

During the study under eight microscope it had already demonstrated that many chloroplasts have a heterogeneous structure made up of small granules called *grana,* which are embedded within the stroma, or matrix. The size of the rana varies from 0.3 to 1.7 μm depending on the species. The smallest ones, within the limit of microscopic visibility, are more numerous. They are flat bodies shaped like platelets or disks, which in a lateral view appear as dense bands perpendicular to the chloroplast surface.

The electron microscope has revealed the true structure of the chloroplast, with its three main components: the envelope, the stroma, and the thylakoids. The envelop is made of a double limiting membrane across which the molecular interchange with the cytosol occurs.

In contrast to the ultrastructure of mitochondria, the inner membrance of mature chloroplasts is not in continuity with the thylakoids (Which could be compared with the mitochondrial crests). In spinach, the outer surface of the envelope os of the order of 500 cm^2 per gm of leaves. Isolated membranes have a yellow colour due to the presence of small amounts of cartenoids. However, these areas lack chlorophyll and cytochromes and contain only 1 to 2 per cent of the protein of the chloroplast.

Stroma fill is a kind of gel-fluid phase that surrounds the thylakoids. This component contains about 50 per cent of the chloroplast proteins and most of these are of the soluble type. It has ribosomes and also DNA, both of which are involved in the synthesis of some of the structural proteins of the chloroplast. As will be discussed later, the stroma is where CO_2 fixation occures and where the synthesis of sugars, starch, fatty acids, and some proteins takes place.

The outer surface of the thylakoid is in contact with the stroma, and its inner surface encloses an *intrathylakoid space.* Thylakoids may be *stacked* like a pile of coins, forming the grana or they may be *unstacked (stroma thylakoids), forming a system of anastomosing tubules that are joined to the* grana thylakoids. The number of thylakoids per granum may vary from a few to 50 or more. The thylakoids contain about 50 per cent of the protein and all the components involved in the essential steps of photosynthesis, such as chlorophyll, the cytochromes, plastocyanine, and others.

By the use of the technique of freez fracturing detailed information about the macromolecular organization of the thylakoid membrane has been gained. By the use of freeze-fracturing, the thylakoid membrane appears as a smooth continuum, representing one half of the lipid bilayer, with paticles of various sizes believed to the protein macromolecules or aggregates which are asymmetrically localized within the lipid matrix.

The freezecleaving of the membrane results in a selective segregation of the particles in either the PF or the EF leaflet. Observation of shows that PFu and PFs contain a large number of small particles, while in EFs the particles are larger and less numerous. Very few of these larger particles are found in the EFu face (corresponding to the unstacked intrathylakoid face). A possible interpretation of these observations will be given later on in this discussion of the chloroplast.

Proplastids are limited by a double membrane. In the presence of light the inner membrane grows and gives off vesicles that arrange themselves to form larger discs. In the granal regions, piles of closely packed thylakoids are built.

In the mature chloroplasts some compartments of the grana remain connected by intergranal tubules. When plants are grown under low light intensity *(etiolation),* the vesicles formed in the proplastid aggregate, forming one or several *prolamellar bodies.* Sometimes the vesicles form a crystalline pattern consisting of regularly connected tubules. When these plants are re-exposed to light, the vesicles may fuse into layers and again develop into grana.

MOLECULAR ORGANIZATION OF THYLAKOID

About the molecular organization of thylakoids, the modern ideas are largely based on the fluid lipid-protein mosaic model of the membrane proposed by Singer and Nicolson in 1972.

This basic concept includes several fundamental notions:

- *Fluidity,* which allows for the free lateral movement of components in the thylakoid membrane;
- *Asymmetry,* in the sense that the crossing of molecules from one half of the bilayer to the other is restricted, permitting the separation of different components;
- *Economy,* since placing the components in a thinmembrane eliminates random movements in the third dimension.

As use have already studied under photosynthesis, in thylakoids there are two *photosystems* (I and II), each consisting of an assembly of many chlorophyll and carotenoid molecules, and a reaction centre where the primary conversion of light into chemical energy takes place. Each assembly of pigments is associated with structural proteins and a system of electron transport to constitute the photosystem. The two photosystems act in series to transfer electrons from water to $NADP^+$. At the same time photophosphorylation is coupled to the electron transport. Here, we will only mention that *photosystem II* (PS II) is involved in the release of O_2 from H_2O, while photosystem I (PS I) plays the major role in the reduction of $NADP^+$.

Lipids represent about 50 per cent of the thylakoid membrane; this includes those directly involved in photosynthesis, such as *chlorophylls* (21 per cent), *carotenoids* (3 per cent). There are also *structural* lipids, which are mainly glycolipids (41 per cent), sulfolipids (4 per cent), and a few phospholipids (10 per cent). Most of the structural lipids are highly unsaturated. The substance chlorophyll is an asymmetrical molecule having a hydrophilic head made of four pyrrole rings bound to each other and forming a porphyrin. This part of the molecule is similar to some animal pigments, such as hemoglobin and cytochromes. In chlorophyll, however, there is a Mg atom forming a complex with the four rings. In animal pigments the Mg is replaced by Fe. Chlorophyll has a long hydrophobic chain (phytol chain) attached to one of the rings.

Regarding higher plants there are two types of chlorophyll—*a* and *b*. In chlorophyll *b* there is a —CHO group in place of the—CH_3 group, indicated by a circle. Three other types of chlorophylls are found in algae, diatoms, dinoflagellates, and in photosynthetic bacteria. Unlike the cytochromes, the chlorophylls are not bound to proteins. A small amount of pigment absorbing at 700 nm has been found. This pigment is called P700 because it is bleached when it absorbs at that wavelength. The pigments concerned with the group called.

Carotenoids are masked by the green colour of chlorophyll. In autumn, the amount of chlorophyll decreases and the other pigments become apparent. These belong to the *carotenes* and *xanthophylis,* which are both related to vitamin.

A Carotenes are characterized chemically by the presence of a short chain of unsaturated hydrocarbon, which makes them completely hydrophobic. Xanthophylls, on the contrary, have several hydroxyl groups.

This is a well known concept that most of the chlorophyll molecules are physically involved with intrinsic proteins, forming chlorophyll-protein complexes. The following complexes have been isolated: the *PS I complex,* containing the reaction centre molecule P700; the *PS II complex,* with the reaction centre molecule P 680; and the *light-harvesting complex.*

Small fragments of membranes rich in PS I and with a high chlorophyll *a*/chlorophyll *b*/ratio have been isolated both from granal and stomal thylakoids; larger fragments rich in PS II and with a low chlorophyll *a*/chlorophyll *b* ratio are derived mainly from granal thylakoids.

LHCP (Light Harvesting Chlorophyll Protein) contains 40 to 60 per cent of the total chlorophyll of the mature thylakoid. This complex is photochemically inactive but transfers light energy to PS II and possibly to PS I, and it contains both types of chlorophyll molecules, It has been proposed that the light-harvesting chlorophylls may be located around their specific intrinsic proteins as shown in the model indicated.

The photophosphorylating system comprises an ATPase similar to that of mitochondria, a *coupling factor* CF_1 located on the outer thylakoid surface, and a CF_0 segment across the membrane. Electron microscopic studies show that CF_1 is localized mainly in stromal thylakoids and is excluded from the stacked regions. The CF_1 can diffuse in the lateral plane of the membrane. Also in this system are the electron transport components of both photosystems. In PS I there is evidence of a vectorial disposition, with the acceptor-site ferredoxin and the ferredoxin $NADP^+$ -reductase situated towards the surface of the membrane.

The localization of the components of PS II is not well established, however, it is likely that this system also has a vectorial arrangement across the membrane. It is generally accepted that both photosystems I and II have their electron acceptors on the outer or stromal side of the thylakoid membrane and that the electron donors are inside (in the thylakoid space), so that the whole electron transport system is situated across the membrane.

PHOTOSYNTHESIS

In plants photosynthesis is one of the most fundamental biological functions. By means of the chlorophyll contained in the chloroplasts green plants trap the energy of sunlight emitted as photons and transform it into chemical energy. This energy is stored in the chemical bonds that are produced during the synthesis of various foodstuffs.

Photosynthesis is somewhat the reverse process. Chloroplasts and mitochondria have many structural and functional similarities, but there are also several differences.

The overall reaction of photosynthesis is:

$$nCO_2 + nH_2O \xrightarrow[\text{chlorophyll}]{\text{light}} (CH_2O)_n + nO_2$$

This shows that, essentially, photosynthesis is the combining of carbon dioxide and water to form various carbohydrates with loss of oxygen.

Each CO_2 molecule from the atmosphere is incorporated into a plant every 200 years, and that all the oxygen in the atmosphere is renewed by plants every 2000 years. Without plants there would be no oxygen in the atmosphere, and life would be almost impossible.

The carbohydrates first formed by photosynthesis are soluble sugars; these can be stored as granules of starch or other polysaccharides inside the chloroplasts or, more usually, inside the amyloplasts. After several steps involving different types of plastids and enzymatic systems, the photosynthesized material is either stored as a reserve product or used as a structural part of the plant (*e.g.*, cellulose).

In reaction H_2O is the hydrogen donor in much the same way that, as shown in the following equation, H_2S is the donor in sulfur bacteria:

$$2H_2S + CO_2 \xrightarrow{\text{light}} (CH_2O) + 2S + H_2O$$

Thus, reaction in higher plants can be written as follows:

$$2nH_2O + nCO_2 \xrightarrow{\text{light}} (CH_2O)n + nH_2O + nO_2$$

Reaction shows that water is the H_2 donor and all the O_2 liberated comes from water. Experiments using water labeled with heavy oxygen ($H_2{}^{18}O$) have confirmed this. In this process water participates primarily as a proton and electron donor. Reaction involves a complex series of steps, of which some take place only in the presence of light and the others take place also in darkness—hence the names *light* and *dark reactions.* In the first, light is absorbed and used by chlorophyll; this is the *photochemical (Hill) reaction.*

In the second or dark reaction CO_2 is fixed and reduced by thermochemical mechanisms.

Under the study of the photochemical reaction, it is necessary to recall the process of oxidatives phosphorylation in mitochondria. In oxidative phosphorylation the flow of electrons is from $NADH_2$ to O_2, following the path of standard oxidation-reduction potentials (*i.e.*, from −0.6 to +0.81 volts); in photosynthesis, the opposite process takes place. The electrons flow from H2O to $NADPH_2$ (*i.e.*, from +0.81 to −0.6 volts).

Regarding photochemical reaction, the primary reactions of photosynthesis, takes place in tylakoid membrane. When these are illuminated, there is a transfer of electrons from water (E'_0 = 0.81 volts).

Transfer against such a negative electrochemical gradient uses the energy provided by the photons of light. Furthermore, the transfer of electrons is accomplished by a chain of *electron carriers* and is coupled with the *phosphorylation* of ADP to ATP. Within the thylakoid membrane, the energy of light is collected by the two *photosystems* PS I and PS II, which act as collector antennae.

Photosystems

There is experimental evidence that the efficiency of photosynthesis is a function of the wavelength of the impinging light.

For example, if chloroplasts are simultaneously excited at 680 and 650 nm, there is considerable stimulation of the photochemical reaction. The existence of two systems of light absorption has been postulated: Photosystem I is stimulated at the longer wavelength and is not accompanied by the production of O_2, whereas photosystem II is activated at the shorter wavelength and yields O_2.

Both photosystems operate in a sequential and interrelated fashion. When photosystem is excited, electrons are boosted to a higher energy level, thereby reducing ferredoxin. This compound is then reoxidized by the transfer of electrons to $NADP^+$, forming $NADPH_2$. Two photons of light (*i.e.*, two excitons) appear to be needed to produce one molecule of $NADPH_2$. This reducing component, having high energy electrons, can be used in a variety of biochemical reactions—particularly in the synthesis of carbohydrates.

To restore the electrons to photosystem I this process of light absorption must be coordinated with photosystem II. In photosystem II the quanta of light remove electrons from the hydrogen of water and boost them to a higher energy level, thus reducing plastoquinone. From here the electrons are brought back to photosystem I by way of an electron transfer system (*i.e.*, cytochrome b559, cytochrome f, and plastocyanine). From this diagram it is evident that only the excited photosystem II results in the release of O_2 from water.

Electron Transport Systems

Freedoxins plays a role in the transport of electrons from photosystem I to $NADP^+$. Ferredoxin is a protein of about 11,600 daltons and contains iron and sulfur but lacks a porphyrin group. Another component is *ferredoxin-NADP oxireductase,* a flavoprotein that uses $NADP^+$ as an electron acceptor. In the transfer of electrons from photosystem II to I there are other carriers. For example, there is *cytochrome* a hydrophobic protein of 100,000 daltons isolated in non-polar solvents; other carriers include the *cytochromes* b_6 and b_3 (or 559), tightly bound to the membrane structure. Chloroplasts also contain a copper-protein called *plastocyanine* and two quinones—*vitamin* K_1 and *plastoquinone*—the latter similar to ubiquinone in mitochondria.

The sequence in the transfer of electrons from photosystem II to I begins with a non-identified carrier Q. This transfer is followed by the sequence: plastoquinone → cytochrome b559 → cytochrome f → plastocyanine. Between the two photosystems the mechanism of interaction is still to be elucidated. Since there is a large pool of plastoquinone molecules, which are lipophilic and very mobile, the link may be not a direct, but a dynamic one.

Photophosphorylation

Regarding mitochondria, in which the electron transport system is coupled with phosphorylation of ADP, in chloroplasts there is a light-induced phosphorylation that is coupled with the electron transfer chain of photosystem II. This process can be uncoupled by 2, 4-dinitrophenol, which also acts in mitochondria.

In mitochondria, the coupling in chloroplasts involves an ATPase having the coupling factor CF_1 on the outer surface of the thylakoid membrane and a CFe hydrophobic segment across the membrane. The mechanisms possibly involved in this coupling are similar to those postulated for mitochondria. Here also, the *chemical-conformational* and the *chemiosmotic* coupling hypotheses have been proposed. The latter, however, is especially favoured in view of the evidence provided by Jagendorf. In his experiments a suspension of spinach chloroplasts with an initial pH of 7 was transferred to a medium with a acid pH 4, they were put in a medium of pH 8, and ADP and Pi were added. Under these conditions there was ATP synthesis driven by the pH difference; this result conforms exactly to the postulate of Mitchell.

In contrast to the oxidative phosphorylation of mitochondria, O_2 is not used in photophosphorylation of chloroplasts. Green plants can produce 30 times as much ATP by photophosphorylation as by oxidative phosphorylation in their own mitochondria. In addition, these plants contain many more chloroplasts than mitochondria. For every one molecule of O_2 liberated eight light quanta are needed. In the overall photosynthetic pathway and two molecules of $NADPH_2$ are formed; the efficiency is close to 25 per cent. Along the electron transport there is the synthesis of three ATP molecules.

The moleculer of ATP and NADPHz produced in the thylakoids provide the chloroplast stroma with the energy needed to fix the CO_2 and to synthesize the carbohydrates. Along with the energy provided by ATP, the reduced $NADPH_2$ can bring about the reduction of atmospheric CO_2 and combine it with the hydrogen to form the various carbohydrates. This process involves many steps, which have been elucidated principally by the use of radioactive CO_2.

The reactions involved are so rapid that they appear one second or less after the addition of CO_2. These reactions occur in complete darkness if the plant has been previously exposed to light.

In cells exposed to CO_2 for five seconds, the dominating compound is 3-phosphoglycerate; Two triose phosphate molecules unite to form hexose (fructose) diphosphate, from which glucose phosphate is formed. Then, from glucose phosphate various disaccharides and polysaccharides are formed.

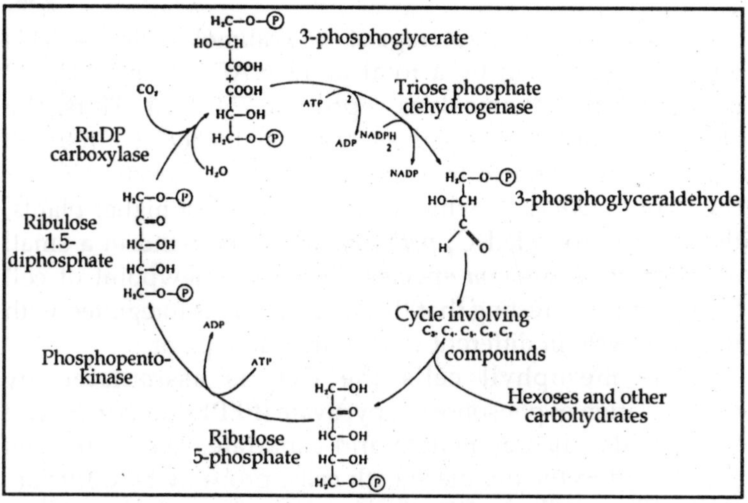

Fig. Diagram of the Calvin or C_3 cycle of photosynthesis in which CO_2 is reduced and carbohydrates are synthesized.

The initial enzyme of the Calvin cycle is *ribulose 1, 5-diphosphate carboxylase (i.e., carboxydismutase)*, which has a large molecular weight (500,000 daltons), eight large subunits synthesized by the chloroplast, and eight small subunits produced in the cytoplasm. This enzyme, which constitutes half the stromal proteins, plays the central roles of integrating into the pentose (ribulose 1, 5-diphosphate) a molecule of CO_2 from the air and of producing two molecules of *3-phosphoglycerate,* by the use of one molecule of H_2O.

The *3-phosphoglycerate* is then phosphorylated by ATP and forms on activated molecule that is able to accept hydrogens and electrons from NADPH. This molecule is *3-phosphoglyceraldehyde,* which is then reduced to from hexose sugars and more complex carbohydrates. Benson and calvin proposed that in this cycle (C_3) a molecule of ribulose 1, 5-diphosphate is regenerated at each turn. Such regeneration is very complex and involves some twelve enzymatic reactions. To produce a hexose from the CO_2 fixation it is believed that six turns of the cycle are needed.

The classical equation for the synthesis of one hexose is:

$$6CO_2 + 12H_2O \xrightarrow{\text{light}} C_6H_{12}O_6 + 6O_2 + 6H_2O$$

This represents the storage of 686 kcal/mole of hexose. This energy is provided by a total of 12 NADPH and 18 ATP molecules that themselves represent 750 kcal. Thus, the efficiency reached is as high as 90 per cent (686/750 × 100 = 90 per cent). It is known that the *photosynthetic carbon reduction* (PCR) or *Calvin cycle,* which is present in most higher plants, there is the so-called C_4 *pathway,* which is found in a small number of *angiosperm* species. From the viewpoint of cell biology it is interesting that this C_4 pathway is integrated with the PCR cycle in different cells of the same plant.

The mesophyll cells the CO_2 is assimilated by carboxylation of phosphoenol pyruvate (PEP), which gives rise to C_4 acids, such as malate and aspartate. These are then transferred to the bundle sheath cells, probably by diffusion. Within these cells the CO_2 that is released by decarboxylation enters the PRC cycle, given origin to 3-phosphoglyceric acid, and the C_3 products may go back to the mesophyll cell to enter the C_4 pathway again.

MODEL OF THE CHLOROPLAST MEMBRANE

The use of biochemical and biophysical techniques, and especially the freeze-fracturing procedure with electron microscopy, support the concept of a fluid mosaic membrane model for the thylakoid membrane. According to this concept, there is a lipid continuum in which protein molecules are partially or totally embeded. As use already know that by solubilizing the membranes with non-ionic detergents, several integral protein complexes had been isolated and characterized. Three of them the *photosystems I and II* and the *light-harvesting complex,* have already been described. In addition, two more complexes have been separated. One corresponds to a *cytochrome f-*b_6 *complex* which, besides the cytochromes, contains non-heme iron, phospholipids, and carotenoids (but not chlorophyll), and which has a minimum molecular weight of 103,000 daltons.

The other complex is the recently isolated hydrophobic protein component of the chloroplast ATPase (CF_0 or HF_0). This component has six major polypeptides with MW between

11,000 and 42,000 daltons. This HF_0 protein binds the CF_1 (coupling factor of ATPase) to reconstitute the enzyme, but is also capable of proton transfer by itself. It should be stressed here that the CF_1, which is analogous to the F_1 of mitochondria, is a peripheral enzyme with a MW near 350,000 and consisting of five polypeptides.

If we try to establish a structural-functional model of the chloroplast membrane, the data obtained by freeze-fracturing and the existence of the above-mentioned intrinsic protein complexes should be considered. The numbers indicate e that the small PF partides are similarly distributed between the unstocked membranes (PFu) or stromal thylakoids and the stacked membranes (PFs), which correspond to the granal partitions. On the other hand, the large particles of the EF leaflet are more concentrated in the stacked region (EFs). Taking into consideration the surface area of the membranes in both regions, it can be calculated that 80 per cent of the EF particles are in the grana, and only 20 per cent in the stromal thylakoids. Several studies tend to identify the large EF particles with photosystem II and the associated light-harvesting complex (the complete PS II complex.

In fact, the proportion of PS II systems in both stacked and unstacked thylakoids is the same as that of the EF particles.. Furthermore, by analyzing the size of EF particles during the development of chloroplasts it was found that while at the beginning both the PF and EF particles are small, with the grinning process, the EF particles increase in size. These data were interpreted as indicating that the light-harvesting complexes were added to the basic "PS II core".

Various integral proton complexes with large particles and a PF leaflet with smaller and more numerous particles. From this model it is evident that while the EF particles can be identified with the complete PS II system, the small PF particles are more difficult to interpret, since they may represent any of the other protein complexes.

Changes in structure of chloroplasts caused by the action of light or darkness have been observed both in vivo and in vitro. In many plant species chloroplasts flatten in the light and become more spherical in the dark. Such changes appear to the mediated by ion fluxes across the membranes of individual thylakoids. In fact, the size of the intrathylakoid space may decrease from 19.5 nm in the dark to 15.2 nm in the light, with a half-time of 1 min. Changes of this type are blocked by drugs that affect ion transport. Similar findings have been observed in vitro in isolated chloroplasts.

When we use freez fracturing it has been demonstrated that during illumination there are changes in the size and density of the particles within the membrane. A more dramatic change was observed in chloroplasts suspended in a solution low in cations. There was unstacking of the membranes coupled with a dispersion of the EF particles, which are normally concentrated in the granal regions. This change is reversible, and upon readdition of cations, stacking is re-established and the EF particles become concentrated in the grana. These observations suggest that the protein complexes have mobility within the fluid membrane.

In the case of chloroplasts the opposite reactions occur; under the action of light H^+ ions are absorbed, and the interior becomes acidified. At the same time K^+ and Mg^{2+} are ejected into the medium. In the dark, H^+ is ejected and K^+ and Mg^{2+} are reabsorbed by the chloroplasts. Thus, the chloroplast membrane has a vectorial organization that is the reverse of the mitochondrial membrane. In physical terms it is understood that dark reactions, being of a themochemical nature, do not need an ordered structural arrangement which are carried out by soluble enzymes. On the contrary, photophosphorylation requires the separation of positive and negative charges into specific pathways of electron flow.

As in mitochondria, as per the *chemiosmotic hypothesis* of Mitchell, the proton gradient may serve as a driving force for the formation of ATP at the CF_1 ATPase.

CHLOROPLASTS AS SEMIAUTONOMOUS ORGANELLES

Mitochondria one of the chloroplasts, exhibit a certain degree of functional autonomy within the intracellular environment. In fact, they undergo division and may contain some genetic information (cytoplasmic inheritance). Chloroplasts contain their own DNA, special ribosomes, smaller than those present in the cytoplasm, and all the necessary molecular machinery to achieve protein synthesis.

Shimper and Meyer accepted in 1883 that plastids multiply by fission. This is easily observed in unicellular algae that contain only one chloroplast. In the alga *Nitella* a division cycle of 18 hours has been recorded cinematographically. The division process is as orderly as chromosomal division. Plastid reproduction by fission implies a growth process of the daughter plastids. (The way in which plastids become differentiated during development and how they change in fine structure under the action of different external factors such as light have been mentioned earlier).

Inheritance does not follow typical Mendelian segregation, thereby indicating an extrachromosomal type of heredity, which has been shown experimentally. When the unicellular green alga *Euglena* is grown in the dark it contains small, colourless proplastids. If under these conditions the nucleus is irradiated with an ultraviolet microbeam, the proplastids are still capable of developing into chloroplasts. However, if the cytoplasm is irradiated with the nucleus shielded, a considerable number of colourless colonies (lacking chloroplasts) develop. These results suggest that there is a cytoplasmic DNA in this alga. It has been generally accepted that a characteristic DNA occurs in chloroplasts of algae and higher plants. DNA regions resembling bacterial nucleoids were identified in thin sections under the electron microscope.

The quantity of DNA per chloroplast varies slightly with the species and is in the range of 2 to 5 × 10^{15} gm, or slightly less than in the bacterium *E. coli.* Segments of DNA as long as 150 μm have been separated from chloroplasts. The replication of chloroplast DNA has been followed with ^{3}H-thymidine. The presence of ribosomes within chloroplasts is easily demonstrated by electron microscopy. These ribosomes are smaller than cytoplasmic ribosomes. Polysomes have also been separated from chloroplasts. The discovery of special ribosomes associated with chloroplasts provided evidence that these organelles contain a specific protein synthesizing system. In the presence of CO_2 as the sole source of carbon, chloroplasts actively incorporate amino acids into proteins. Protein synthesis is preferentially inhibited by chloramphenicol concentrations that do not affect protein synthesis in the cytoplasm. The involvement of the two types of protein synthesis in the assembly of chloroplasts may be studied by using cycloheximide to inhibit synthesis due to cytoplasmic ribosomes.

Chloroplasts divide, grow, and differentiate; they contain DNA, ribosomal RNA, and messenger RNA, and are able to conduct protein synthesis. It has been suggested that chloroplasts may have resulted from a symbiotic relationship between an autotrophic microorganism, one able to transform energy from light, and a heterotrophic host cell. However, it is evident that chloroplasts have in their DNA the genetic information to code for only a limited number of proteins and the elements necessary to establish a DNA-RNA-directed protein synthesis (*i.e.*, ribosomes, tRNA, and so forth).

Inchloroplasts protein may be synthesized:

- By an exclusive chloroplastic mechanism,
- By a mechanism involving nuclear genes and chloroplastic ribosomes,
- By nuclear genes and cytoplasmic ribosomes. Although the symbiont hypotheses is attractive, it is evident that synthesis and assembly of the components integrating a chloroplast depend to a great extent on nuclear genes and on cytoplasmically directed protein synthesis.

7

Cell Division

Cell division is the complex phenomenon by which cellular material is divided equally between daughter cells. This process is the final, and microscopically visible, phase of an underlying change that has occurred at molecular and biochemical levels. Before the cell divides by mitosis, its fundamental components have duplicated-particularly those involved in hereditary transmission. In this respect, cell division can be considered as the final separation. of the already duplicated macromolecular units.

Special consideration will be given to the *mitotic spindle, the* fibrous component made of microtubules which is assembled every time the cell begins to divide and is disassembled at the end of mitosis. It will be seen that the mitotic spindle functions as a structural framework and force generating system the sets the chromosomes in position and distributes them between the daughter cells. In our discussion of mitosis we will also deal with some of the structural differentiations of chromosomes, such as the *kintochores* and *cenntromeres* to which the spindle microtubules are annfchored, and the general changes in fine structure that chromosomes undergo during the mitotic cycle.

MITOSIS—A GENERAL DESCRIPTION

In spite of the fact the overall process of mitosis is similar in all eukaryotic cells, there are variations between different organisms—particularly between animal and plant cells—that make it difficult to give a general description. For example, it will be mentioned later that in higher plant the spindle has no *centrioles or asters* at the poles and furthermore, that the separation of the daughter cells is a more complex phenomenon than in animal cells. Although plant *meristems* (*e.g.*, young root tips) are particularly advantageous for the study of celly division, since 10 to 15 per cent of the cells are mitosis our description will be based mainly on animal cells such as those in tissue culture or dividing eggs.

PROPHASES

The beginning of *prophase* shown by the appearance of the chromosomes as thin threads inside the nucleus. In fact the word "mitosis" is an expression of this phenomenon, which becomes more evident as the chromosomes start to condense. The condensation occurs by a process of coiling or folding of the chromatin fibres. At the same time the cell becomes spheroid more refractile and viscous.

Each prophase chromosomes is composed of two coiled filament, the *chromatids,* which are a result of the replication of the DNA during the S period. As the prophase progresses, the chromatids become shorter and thicker, and the primary constrictions, which contain the centromeres and kinetochores, become clearly visible During early prophase, the chromosomes are evenly distributed in the nuclear cavity; as prophase progresses the chromosomes approach the nuclear envelops to become empty. The centrifugal movement of the chromosomes indicates that the disintegration of the nuclear envelope is approaching and with it the end of prophase.

At this time each chromosome appears to be composed of two cylindrical, parallel elements that are in close proximity. Not only the primary constrictions, but also the secondary constrictions along some chromosomes may be observed Other changes taking place within the nucleus are the reduction in size of the nucleoli, and their disintegration within the nucleoplasm. At the end of prophase, with the rapid fragmentation and disappearance of the nuclear envelope the nucleolar material is released into the cytoplasm.

While these processes are taking place in the nucleus, in the cytoplasm the most conspicuous change is the formation of the spindle. In early prophase there are two pairs of centrioles, each one surrounded by the so-called *aster*, composed of microtubules that radiate in all directions. (The name "aster" refers to the starlike aspect of this structure). The two pairs of centrioles migrate along with the asters, describing a circular path towards the poles, while the spindle lengthens between. The migration of the asters continues until they become situated in antipodal positions.

Centrioles replicate during interphase, generally during the S period. At the beginning of prophase there is a single aster surrounding the two pairs of centrioles One to the pairs remains in position, with half the original aster, while the other pair along with the other half aster, migrates about 180° around the periphery of the nucleus to reach the opposite pole.

Sometimes the transition between prophase and metaphase is called *prometaphase.* This is a very short period in which the nuclear envelope disintegrates and the chromosomes are in apparent disorder. After that the spindle fibbers invade the central area and their microtubules extend between the poles. The chromosomes become attached by the kinetochrores to some of the spindle fibres, and they undergo oscillatory movements until they become radially oriented in the equatorial plane and form the *equatorial plate* Those fibbers of the spindle that connect to the chromosomes are called the *chromosomal fibres;* those that extend without interruption from one pole to the other are the *continuous fibres.*

Mitosis in which the spindle has centrioles and asters is called *astral* or *amphiastral* and is found in animal cells and some lower plants. Mitosis in which centrioles and asters are absent is called *anastral* and is found in higher plants, including all angiosparms and most sperms. Centrioles and asters are not indispensable to the formation of the spindle. In a certain way, in *astral mitosis* the formation of the spindle is a mechanism which leads to the distribution of the centrioles between the two daughter cells.

At anaphase the equilibrium of forces that characterizes metaphase is broken by the separation of the centromers —a process which is carried out simultaneously in all the chromosomes. The centromers move apart, and the chromatids separate and begin the migration towards the poles. The centromere always leads the rest of the chromatid or *daughter chromosome,* as if this is being pulled by the chromosomal fibres of the spindle.

The chromosome may assume the shape of a V with equal arms, if it or with unequal armas, if it is metacentric, or with unequal arms if it is submetacentric. During anaphase the microtubules of the of the chromosomal fibres of the spindle shorten one third to one fifth of the original length. Simultaneously, the microtubules of the continuous fibres increase in length. Some of these stretched spindle fibres now constitute the so-called *interzonal fibres.*

The end of the polar migration of the daughter chromosomes marks the beginning of telophase. The chromosomes start to uncoil and become less and less condensed by a process that in some ways, recapitulates prophase, but in the reverse direction. At the same time, the chromosome gather into masses of chromatin which become surrounded by discontinuous segments of nuclear envelope. Such segments fuse to make the two complete nuclear envelopes of the daughter nuclei during the final stages the nucleoli reapear at the sites of the nucleolar organizers. Simultaneous with the uncolling of the chromosomes and the formation of the nuclear envelope, *cytokinesis* occurs. This is the cytoplasm. In animal cells the cytoplasm constricts in the equatorial orgion, and this constriction is accentuated and deepened until the cell divieds.

In cells of higher animals the period of cytokinesis is marked by active movements at the cell surface that are best described as "bubbling." Some investigators suggest that bubbling may reflect the activity of a rapidly expanding membrane. In amoeboid cells at telophase both daughter cells have active movements, which appear to pull them apart. This is best observed in films of dividing cells and in scanning electron micrographs. The high viscosity of the cytoplasm, characteristic of metaphase and anaphase, decreases during telophase. In most cases asters become reduced and tend to disappear. During cytokinesis the cytoplasmic components and the Golgi complex.

Function Role of the Mitotic Apparatus and Molecular Organization

The term mitotic apparatus has been applied to the *asters* that surround the centrioles and to the *mitotic spindle.* In fixed preparations the aster appears as group of readiating fibres that converge towards the centrioles. Around the centrioles there is often a clear zone called the microcentrum of centrosome into which the fibres of the aster do not penetrate.

The spindle has the so-called chromosomal fibres, joining the chromosomes to the poles; the continuous fibres, extending pole to plole; and the interzsonal fibre observed between the daughter chromosome and cuclei in anaphase and telophase. All these fibres, including those of the aster, are composed of microtubules.

The fine structure of the mitotic apparatus is studied principally with the electron microscope. Interesting observations may also be made using polarization microscopy and the Nomarski interference-contrast microscope. Because of their positive bire-fringence spindle and aster fibres are readily observed in living cells with the polarization microscope. In plant which are devoid of centrioles and asters, the first spindle fibres appear at prophase in a clear zone surrounding the nucleus Bire-fringence is strongest near the kinetochores but becomes weaker towards the ples. During anaphase, the chromosome are led by intensely birefringent chromosomal spindle fibres The continuous fibres. in which Bire-fringence is very low in early anaphase. become more conspicuous in late anaphase and telophase. In animal cells such tibers form a kind of buridle that maintains a connection between the two daughter cells for some time. Under the electron microscope this bundle of microtubules appears surrounded by a dense material and some vesicles that together constitute a structure called the *midbody*.

The use of antibodies against tubulin (the main component of microtubules) has permitted interesing observations on the changes in the mitosis in cultured cells. The degree of fluorescence is related to the number and condensation of the microtubules in the aster and in the various regiouns of the spindle.

In the basic constriction there is a special differentiation of the chromosome called the *centromere,* which is associated with a proteinaceous structure, the *kinetochore. The* kinetochore is the site of implantation of the microtubules that constitute the chromosomal spindle fibres; for this reason it has been considered as functionally related to the chromosomal movements during mitosis. Usually two kinetochores one in each chromatied, are observed but in some cases a large number may be found *(diffuse centromeres).* In these instances, the microtubules are implanted at many points along the chromosomal arems. Diffuse kinetochores have been found in groups of plants *(e.g., Luzula).* in insects *(e.g., Hemiptera),* and in some algae. The kinetochore may be detected during prophase, before there is a connection with the microtubules. Under the electron microscope the kinetochore appears as a plate or cup-like disc, 0.20 to 0.25 μm in diameter, plastered upon the primary constriction.

In a cross section the kinetochore consists of the following three layers and *outer layer* of lower density, an *inner* dense layer in contact with the underlying chromatin fibres. Emanating from the convex surface of the outer layer, in addition to the microtubules, a "corona" of fine filaments has been observed.

The propers chemical nature of the kinetochore is unknown, although it is thought that it represents a non-chromation material added to the surface of the primary constriction. It has been suggested that the kinetochore is a gene product and that its activity is genetically controlled. [7] At present there is no evidence that the centromere itself contains genetic material.

The main function of kinetochores seems to be related to the attachment of the chromosomal microtubule. A related function may be to serve as nucleation centres for the polymerization of tubulin. Studies at the electron microscopic level using anti-tubulin antibodies have revealed that tubulin is found not only in the spindle and centrioles (both of which contain microtubules)but also in kinetochores. This suggests that they contain the nucleating intermediates necessary for the assembly of tubulin monomers into microtubules.

There 16 microtubules in the spindle of yeast cells and as many as 5000 in the spindle of a higher plant cell. A precise study of the number of microtubules has been carried out in dividing animal cells in the various regions of the spindle.

In a cultured cell line it has been found that about 34-5 microtubules end at each kietochrore. These *kinetochore microtubules* correspond to the chromosomal fibres mentioned in the older literature. The microtubules point towards the poles but not all of them are long enough to reach the pole. Quantitative measurements have also shown that the so-called continuous fibres are not actually so, since only a few microtubules may be so long enough to reach the poles.

As better names, in addition to kinetochore microtubules, the terms *polar* and *free* microtubules have been proposed. Several reports indicate that microtubules may have some kind of cross-bridges which may hold the spindle together. Fine microfilaments have been observed in between the microtubules which may correspond to actin. The presence of actin in parts of the spindle, especially near the poles, has been demonstrated by anti-actin antibodies. However, microfilaments do not, appear to be connected to the kinetochores. The spindle also contains ribonucleoprotein particles presumably derived from the nucleolus, small membrane vesicles, and a fine fibrous matrix.

The cyclic changes in the bire-fringence of the spindle are interpreted as reflecting the systematic assembly and disassembly of the microtubules. Studies employing micromanipulation have shown that, within certain limits, the spindle fibres resist extension, maintain mechanical integrity, and are instrumental in the movement of the chromosomes.

Studies living cells also reveal that the spindle fibres represent a very dynamic structure. Their bire-fringence is abolished in a matter of seconds by low temperature, but after return to normal temperature, but after return to normal temperature, the cell recovers in a few mitosis. with continuation of the arrested mitosis, Intense hydrostatic pressure, microbeam ultraviolet irradiation, and certain drugs, such as colchicine Colemid, and others also induce disappearance of the bire-fringence and the microtubules. One interesting change is produced with heavy water. When dividing sea searching oöcytes are placed in 45 per cent D_2O, the bire-fringence increases twofold, and the volume of the spindle increases about tenfold in 1 to 2 minutes. After the egges are returned to H_2O the birefringence reverts to normal within a few minutes· These experiments imply that in cytoplasm there is a great excess of buiding material bor the mitotic apparateus.

By using gentle methods of cell fractionation on sea urchin eggs, Mazia and Dan made the first isolation of the mitotic apparatus. Slow progress was made throughout the years in separating this labile structure from various cells.

A new approach has recently been used based on the discovery of the conditions under which brain tubulin could be polymerized in vitro. The technique consists of preserving the isolated spindle in a medium containing exogenous tubulin. With this method, some the spindle functions are preserved. For example, in calls lysed at anaphase the chromosomes have been observed to move in normal fashion.

Control of the Assembly

Some of above-mentioned observations suggest that the assembly of microtubules is *controlled* by the *poles* and also by the *kinetochores*. A third mechanism may involve the *lateral interaction* between the spindle microtubules. Although the exact mechanism is unknown, it is though that it involves first the phlymerization of tubulin subunits and then their assembly into microtubules by secondary bonds formation.

Normally in a cell there is a large *pool* of *tubulin monomers* which are in a kind of dynamic equilibrium with the polymerized microtubule. An equilibrium is also established with the *oytoplasmic microtubules*. When cells enter prophase, these microtubules become depolymerized and are replaced by the mitotic spindle. At metaphase, only the spindle microtubules are present; at anaphase, with the movement of the chromosomes, the spindle become depolyemrized, and at telophase the daughter cells are held by the *midbody* and the cytoplasmic microtubules reappear.

Action of Ca^{2+}

The tinding that Ca^{2+} can inhibit the polymerization of tubulin has led to the suggestion that the intracellular concentration of Ca^{2+} may play an important role in the assembly and disassembly of microtubules in vivo. The level of Ca^{2+} in the cytosol may be regulated through various mechanisms by which this cation is bound or sequestered in the cell. Current hypotheses about the role of the mitotic apparatus are based on the idea that the microtubules can generate some sort of mechanical force either by "pushing" or by "pulling" the other cell components. These two mechanisms involve the elongation of or the shortening of the microtubules.

In prophase, the migration of centrioles towards the poles is probably caused by the "pushing" which is a result of the elongation of continuous fibres. In fibroblasts separation of centinuous fibres. In fibroblasts, separation of centrioles occurs at a rate of 0.8 to 2.4 μm per minute. As soon as the nuclear envelope begins to disintegrate, the nuclear region is invaded by microtubule that establish pole-kinetochore attachment before the chromosomes move towards the metaphasic plate.

We have already studied the poles of the spindle and the kinetochore are the two main nucleation centres of the spindle microtubules. The motion of chromosomes towards the metaphase plant can be explained as the result of lateral interaction between the microtubules arising from the kinetochore and from the pole microtubules.

During metaphase an equilibrium of forces is established. There are pole-directed forces that tend to pull the sister kinetochores towards the poles; however, since these are linked until the onset off anapase, the opposing forces cancel each other. It is possible that the pole microtubules could exert a counteracting force that would tend to pull the poles inward (Such an equilibrium of forces could also explain the bowed shape of the spindle and the fact that the chromosome remain stationary at metaphase. With the separation of the sister chromatids at anaphase, this equilibrium is broken. The chromosome usually move towards the poles with a rate of about 1μm per min.

The characteristic shapes assumed by the chromosomes in metaphase and anaphase and the anaphase bridges that may result from the stretching of dicentric chromosomes suggest that the forest responsible for the pulling of the chromosome towards the poles is transmitted to the kinetochores These forces may be great enough to produce the rupture of these dicentric chromosomes. It has been calculated that to move a chromosome a force of about 10^{-8} dynes is needed, and that the entire displacement—from the equator to the pole of a chromosome fay require the use of about 30 ATP molecules.

The chromosomes move at a constant rate until they are separated by a distance equivalent to the length of the metaphase spindle; then they slow down. It is important to remember that during anaphase the spindle elongates until in nearly duplicates the length that it had at metaphase with a shortening of the kinetochore microtubules, which implies that there are pole-directed forces acting on the kinetochrores. Recently a better view of the spindle microtubules at different stages of mitosis has been obtained in thick section of cultured cells by high-voltage electron microscopy (100 kvolts).

The resultant images suggest that at anaphase there are two distinct events:

1. The motion of chromosome towards the poles on each half spindle and
2. The separation of the free microtubules at the interzonal region. At the midregion of the interzonal fibres it is possible to identify a dense zone classically known as the ***stem body***. Electron microscopic observations have shown that in this region there is a high concentration of interdigitating microtubules.

In recent theories it has been to explain the role of the mitotic apparatus in mitosis have been influenced by the discovery of a cyclic mechanism in muscle and cilia in which there is a direct coupling between the hydrolysis of ATP and mechanical events Such a coupling is based on an elaborate and rather stable molecular organization of these systems. On the other hand, the mitotic apparatus is very labile and is assembled and disassembled at every mitotic cycle.

Two main hypotheses have attempted to interpret the molecular mechanism responsible for the movement of chromosomes:

1. The dynamic equilibrium hypothesis, and
2. The sliding mechanism hypothesis.

Two dynamic equilibrium hypothesis first proposed in 19 by Ostergren has been established mainly by Inoue and Sato as a consequence of their studies with the polarization microscope.

This hypothesis postulates that the polymerization-depolymerization of microtubules is directly responsible for the movement of the chromosomes. While it is rather easy to see that the microtubules could have "cytoskeletal" role, serving as a kind of guide guide for maintaining the shape of the cell and for "pushing" the centrioles, it is more difficult to understand haw its depolymerization could provide a motive forces that However, calculations of the cohesive forces that could be involved in assembling microtubules have led to the conclusion that they contain enough energy to move chromosomes through the cytoplasm.

The *sliding bypothesis,* although recognizing that spindle microtubules assemble and disassemble does not admit that they generate directly the driving force to move the chromosomes Rather, they are just the "railroad tracks" on which the chromosome move. This hypothesis postulates the involvement of lateral interactions be way of molecules that could play a role analogous to *dynein* in flagellate or by interaction with other bibbrous components such as *actin*. We mentioned earlier than actin may be demonstrated in certain parts of the spindle.

Furthermore, the use of antibodies against myosin has shown that *myosin* may be present in the spindle. Thus, it possible, *although the issue is not yet settled,* that the interaction of spindle catin with myosin could provide the motive force in chromosomal movement, and that the microtubules could play a mechanical case of *cell motility.* A molecular mechanism for the function of the mitotic apparatus still seems to be long way ahead.

The separation of the daughter nuclei and *cytokinesis* or *cell cleavage* may be two separate processes. For example, the eggs of most insects undergo division of the nucleus to form a multinucleate *plasmodium* without separation of the cytoplasmic territories. Cleavage of animal cells has been studied mainly in tissue culture and dividing eggs and is produced by the furrowing of the cell. The first visible changes consist of the appearance of a dense material around the microtubules at the equator of the spindle at either mid or late anaphase.

Then although spindle microtubules tend to disorganize and disappear during telophase, they usually persist, and may even increase in number at the equator, frequently being intermingled with a row of vesicles and the dense material; the entire structure is called the *midbody* Simultaneously, there is a depression on the cell surface—a kind of constriction that deepens gradually until reaching the midbody. With the completion of the furrowing the separation of the cell is concluded. The importance of the spindle of the study of multipolar cell divisions in which a furrow is formed between each pair of asters. Even the removal of the whole mitotic apparatus of the sea urchin egg, however, does not inhibit cell cleavage.

Cell cleavage by furrowing may be the result of a contractilering mechanism that involves the cell cortex. A contractile protein exhibiting. ATPase activity and properties similar to myosin was isolated from dividing sea urchin eggs. These experiments should be correlated with other results showing that ATP may produce cytoplasmic contraction in different cell types. A system for fine microfilaments may be directed under the plasma membrane at the region of the furrow. It can be demonstrated that such filaments are composed of an actin-like protein. Another indication that a contractile mechanism is acting in cell cleavage is provided by the action of the drug *cytochalasin,* which is able to inhibit such a process as well as other contractile mechanismes in cells.

Studies using antibodies against *actin* and *myosin* have shown that both these proteins are present in the contractile ring of cytokinesis. In plant cells at anaphase the interzonal region of the spindle become transformed into the *phragmoplast,* which in some way is similar to the midbody of animal cells.

Under the electron microscope it is possible to observe that the interzonal microtubules at the equator plane have scattered patches of vesicles and of dense material applied to their surface. The vesicles are derived from the Golgi complexes which are found in the regions adjacent to the phragmoplast and which migrate into the equatorial region to be clustered around the microtubules.

Although the phragmoplast initially is found as a ring in the periphery of the cell, with time it grows centripetally by addition of microtubules and vesicles until it extends across the entire equatorial plane. The vesicles increase in size and then fuse until the two daughter cells become separated by fairly continuous plasma membranes. At this time the phragmoplast has been transformed into the *cell plate.*

Thin cytoplasmic connections—the *plasmodesmata*—traverse the cell plate and remain in place for communication between the adjacent daughter cells. The making of the cell plate also leads to the synthesis of the *cell wall.* The *Golgi* vesicles in the phragmoplast are already filled with a secretory material consisting mainly *pectin.* The fusion of the vesicles results in the combining of the pectin in the extracellular space between the daughter cells, thereby forming the main body of the *primary cell wall.* Later on, microfibrils of cellulose are laid down in a semicrystalline lattice on the two surfaces of the facing daughter cells.

8

Cell Reproduction

Meiosis is a special type of cell division that is found in organisms in which there is *sexual reproduction*. In many protozoa, algae, and fungi reproduction is asexual, i. e, by simple cell division or *mitosis. In* this form of reproduction all the descendant individuals have *uniparental inheritance.* On the other hand, in most multicellular organisms—animals or plants—the reproduction is sexual. This means that during their lifetime they produce *sexual cells* or *gametes (eggs* and *sperms)* which by their fusion in the *zytote* will originate the new organism. It is evident that in this process there is a *biparental inheritance* which implies a process of nuclear fusion called *fertilization*.

To understand the essentials of meiosis let us take the case of man mitosis were the only type of cell division, then each gamete would have 46 chromosomes and the zygote. This would be repeated in the next generation leading to an increase according to the square of the number of generations! *Meiosis* is the mechanism that prevents this from happening. By a series of two divisions, the number of chromosomes is reduced by half, which gives rise to four *haploid* cells (four spermatozoa in the male, and one ovum and three polar bodies in the female). The processes by which the gametes are produced are called *spermatogenesis* and *oögenesis* respectively, and they take place in the gonads (*i.e.,* testicle and ovary).

The type of meiosis found in man is characteristic of all animals and of a few lower plants, and it is called *terminal* or *gametic,* because the meiotic divisions occur just before the formation of the gametes. We shall see later that in most plants *meiosis* is *intermediary* or *sporic* and occurs at some time between fertilization and the formation of gametes. In other words, in this case meiosis is preceded and followed by mitosis. Cells undergoing meiosis are often called *meiocytes.*

Comparison between Mitosis and Meiosis

Many of the events that we have studied in mitosis are also found in meiosis, for example, the sequence of changes in the nucleus and cytoplasm; the stages of prophase, metaphase, anaphase and telophase; the formation of the mitotic apparatus with the asters and spindle; the condensation cycle of the chromosomes; and the structure and function of the centromeres.

There are, however, essential differences between the five proceses, some of which are the following:

1. Mitosis occurs in all *somatic cells* of an individual, but meiosis is limited to only the *germinal cells.*
2. In mitosis each replication cycle of DNA is followed by one cell division. The resulting daughter cells have *diploid number* of chromosomes and the same amount of DNA as the parent cell. In meiosis one replication cycle of DNA is followed by two divisions, and the four cells are *haploid* and contain half the amount of DNA.
3. In mitosis DAN synthesis occurs in the S period, which is followed by a G_2 phase before the onset of division. In meiosis there is *premeiotic* DNA synthesis which is much longer than that in mitosis and which is followed immediately by meiosis. In other words, the G_2 phase is short of non-existent.
4. In mitosis every chromosome behaves independently; in meiosis the *homologous chromosomes* become mechanically related during the first mechanically related during the first meiotic division. (*i.e., meiotic pairing).* While mitosis is rather brief (one or two hours), *meiosis is a long process.* For example, in the human male it may last 24 days, and in the female it may go on for several years.
5. A fundamental difference between the two types of

cell division is that in mitosis the genetic material remains constant (*i.e.,* with only rare mutations or chromosomal aberrations,) while *genetic variability* is one of main consequences of meiosis.

MEIOSIS

Several mitotic divisions of the spermatogonia (or oögonia), meiotic division starts A kind of switch shifts these cells from mitosis to meiosis. In the G_2 period of interphase there is apparently a decisive change that directs the cell towards meiosis instead of towards mitosis Some experiments in cultured cells of lily anthers show that this change probably takes place at the beginning of G_2 but its exact nature is still unknown. In these cells it is possible to experimentally reverse the switch back to mitosis before the process becomes irreversible.

- MEIOSIS
  - DIVISION I
    - Prophase I
      - Preleptonema
      - Leptonema
      - Zygonema
      - Pachynema
      - Diplonema
      - Diakinesis
    - Prometaphase I
    - Metaphase I
    - Anaphase I
    - Telophase I
  - Interphase
  - DIVISION II
    - Prophase II
    - Metaphase II
    - Anaphase II
    - Telophase II

We shall see that the following are the essential processes of meiosis:

1. The *pairing (i.e., synapsis) of* the homologous chromosomes,
2. The formation of *chiasmata,* which represent the underlying *genetic recombination* (*i.e.,* the interchange of genetic material),
3. The *segregation of the homologous chromosomes.*

Form the morphologic viewpoint the first meiotic division is characterized by a long prophase during which homologous chromosomes pair closely and interchange hereditary material. The classic stages of mitosis do not suffice to describe the complex movements of the chromosomes in meiosis. The successive meiotic stages are the following:

Leptotene Chromosomes always Appear to be Single and Have Chromomeres

Preleptonema corresponds to the early prophase of meiosis. Chromosomes are extremely thin and difficult to observe Only the sex chromosomes may stand out as compact heteropyknotic bodies

Preleptonema (G.r., *leptos,* thin+*nema,* thread) corresponds to a period in which the nucleus has increased in size and the chromosomes have become more apparent. The leptotene chromosomes differ in the following two ways from the prophase mitotic ones. In spite of the fact that they have two chromatids leptotene chromosomes look single rather than double. These chromosomes show bead-like thickenings, the so-called- *chromomeres,* which appear at irregular intervals along their length. Since these beads are charateristic in size, number and position for a particular chromosomes they may be used as landmarks to identify a specific chromosomes of an organism, With the further contraction of chromosomes during zygonema and pachynema, chromosomes become larger and fewer in number. Under the electron microscope the chromosomes show the unit fibre flooded back and forth at the chromomeres.

Chromomeres were once thought to represent the genes. However, they are too few in number (1500 to 2500 in the lily) to include all the genes. It is interesting to mention at this point that during the entire meiotic prophase (and also during the mitotic) the chromosomes become progressively shorter without becoming entangled, which suggests that during interphase the DNA stands are already in a three-dimensional ordered pattern. Frequently leptonemic chromosomes have a definite polarization and form loops whose ends are attached to the nuclear envelope at points near the centrioles, contained within an aster. This peculiar arrangement is often called the "bouquet.

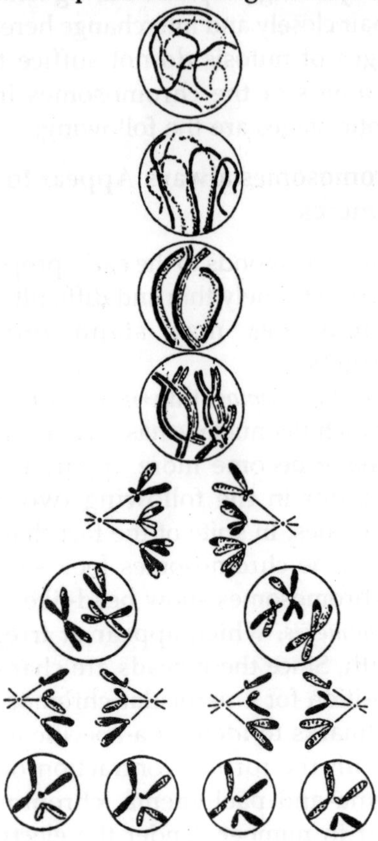

Fig. General Diagram of Meiosis, Illustrating the Union, Separation, and Distribution of the Chromosomes

Synapsis of the Chromosomes

During *zygonema,* the first essential phenomenon of meiosis occurs. The homologous chromosome become aligned and undergo pairing in ap process often called *synapsis of the chromosomes.* Pairing is highly special structure which is observed under the electron microscope and is generally named the *synaptonemal complex.*

It was first described by Moses in 1956 this complex is composed of two *lateral components* or arms and a *central* or *medial* element which are both interposed in between the two pairing home logues. The synaptonemal complex can be considered is the structural basts for pairing and synapsts of meiotic chromosomes. It is important to emphasize that at the end of lep-tonema the lateral elements of the SC have already appeared in the space between the two chromatids, while the medial component appears, with the pairing at zygonema.

Another important point is that when pairing starts the packing of the chromosomes is already very high on the order of at least 300/1 (*i.e.*, the DNA is 300 times longer than the length of the chromosomes. In other words, only 0.3 per cent of the DNA of the homologous chromosomes is matched along the length of the lateral component of the SC. To explain this it is generally assumed that the unit fibre of the chromosome makes loops which at one point joint the SC. These points may function as matching sties for the synaptic alignment of the chromosomes. It is supposed that the prezygotene condensation is a highly ordered process and that it follows a similar pattern in both homologues, so that the synapsis becomes specific for the corresponding set of genes.

During zygonema the homologous chromosomes become aligned. This is made easier by the fact that the telomeres of the chromosomes are frequently inserted in the nuclear envelope.

The paring does not hove a special starting point. Sometimes the chromosomes unite at their polarized ends and continue pairing at the antipodal extremity; in other cases, fusion occurs simultaneously at various places along the length of the thread. Polarization seems to favour regularity in pairing. Pairing is remarkably exact and specific; it takes place point for point, and chromomere for chromomere, in each homologue. The two homologues do not fuse during pairing but remain separated by a space of about 0.15 to 0.2 μm, which is occupied by the synaptonemal complex.

The morphology of the SC is very similar in plant and animal meiocytes. It has a tripartite structure, with, the two lateral arms in each homologous chromogous chromosome and the central component. In cross section it can be observed that the SC is a flattened, ribbon-like structure The lateral arms vary in width from 20 to 80 nm in various species. They are formed of electron-dense coarse granules or fibres. These arms are joined to the adjacent chromosomes by fine fibrils.

In most plants and animals—with the exception of the insects—the two side arms are separated by an axial space of lowerdendity. In insects the central components may be very complex. In the case it has the aspect of a ladder, with three dense parallel lines and bridges crossing at intervals of 20 to 30 nm. These bridges are formed by fine fibrils that span the central and lateral components and are arranged perpendicularly to them. By the use of serial sections it has been possible to reconstruct all the bivalent chromosomes of an oocyte. Recently a special squashing technique has permitted the direct observation of the SC of all the bivalents in the karyotype. In the human spermatocyte it has been found that the SC is present only in the region of pairing between the lateral components of the X and Y chromosomes.

Researchers employing deoxyribonuclease to digest DNA, and the heavy, metal indium to stain nucleic acids have suggested that the main component of the SC is protein in nature. This protein is basic and probably similar to the histones. The fine fibres that cross between the two lateral arms and connect with the chromosomes probably contain.

Pachynema—The pairing of Chromatids

During pachynema (G.r., *pachus* thick), the pairing of the chromosomes reaches completion. The chromosomes contract longitudinally, resulting in shorter and thicker threads. By middle pachynema, the nucleus contains half the number of chromosome; Each unit as a *bivalent* or *tetrad* which is composed of two homologous chromosomes in close longitudinal union and which contains four chromatids. Each homologous chromosome has an independent centromere; however, under the electron microscope it has been observed that each chromatid has its own centromere.

Thus, in a tetrad there are four centromeres, two homologous and two sister. However, during the first meiotic division the centromere of the two chromatids behave as a functional unit. In late pachytene a line of separation perpendicular to the plane of pairing appears, and the four chromatids become visible. The chromatids of each homologue are called *sister chromatids*. During pachynema, the space occupied by the SC is maintained, and all the homologous chromosomes have finished the pairing process.

Experimental evidence suggests that during pachynema two of the chromatids of the homologues exchange segments (i. e., recombine) It is thought that transverse breaks occur at the same level on each of the chromatids and that this event is followed by interchange and final fusion of the chromatid segments. Pachynema is usually a long-lasting stage of prophase, whereas leptonema and zygonema may last only a few hour. Pachynema may last for days, weeks, or even years.

Pairing of Homoeogues

We have already mentioned that the lateral arms of the SC appear in leptonema, before pairing, and that completion of the SC occurs during zygonema and becomes more conspicuous at pachynema. After this stage, during diplonema, the SC disintegrates and usually disappears. One function of the SC appears to be that of stabilizing the pairing of the homologues, another, that of facilitating recombination, Because of the deposition of new protein molecules in the lateral arms of the SC the matching segments of DNA are placed in a way that allows interchange at the molecular level. The SC can be interpreted as a protein framework that permits only the proper alignment of the homologues but also their recombination.

It must be assumed that the DNA fibres of the paired chromatids are within 1.0 nm of the central component of the SC in order for breakage and recombination to take place. This could be achieved by the fine DNA fibrils that are arranged like transverse bridges and that connect the homologous chromatids with the lateral and central elements of the SC. It is thought that the homologous sequences or nucleotides search for each other and achieve a close pairing inside the central component of the SC. It is within these special regions that recombination, with the exchange of DNA segments, occurs. Electron microscopic studies of meiosis at the stage of pachytene have revealed the presence of dense nodules of about 100 nm, in intimate association with the synaptonemal complex. The number of these nodules and their location along the bivalents is related to the number and distributions of the genetic exchanges (crossovers). The name *recombination nodule* has been applied to these, with the suggestion that they may be involved in the recombination process taking place at pachytene of meiosis.

At *diplonema* the intimately paired chromosomes repel each other and begin to separate *(E)* However, this separation is not complete, since the homologous chromosomes remain united by their points of interchange or *chiasmata*. Chiasmata are generally regarded as the expression of the phenomenon called *crossing over*, or *recombination*, by which chromosomal segments with blocks of genes are exchanged between homologous members of the pair.

With few exceptions chiasmata are found in all plants and animals. At least one chiasma is formed for each bivalent. Their number is variable, since some chromosomes have several. During diplonema the four chromatids of the tetrad become visible; under the electron microscope the synaptonemal complex can no longer be observed. Diplonema is long-lasting period. In the fifth month of prenatal life, for example, human oöcytes have reached the stage of diplonema and remain in it until many years later, when ovulation occurs. In most species there is uncoling of the chromosomes during diplonem. In fish, amphibian, reptilian and avian ooytes, however, the uncoling becomes so marked that the greatly enlarged nucleus assumes an interphase appearance.

In these cases the bivalent chromosomes may strain a special configuration known as the *lampbrush chromosome*, in which their chromonema uncolis into loops that converge upon a more cold axis. It will be shown later that the presence of these lampbrush chromosomes is related to an intensive RNA synthesis and to the enormous growth of the oocyte.

Diakinesis and Chiasmata

In diaclasis the contraction of the chromosomes again becomes accentuated. The tetrads are more evenly distributed in the nucleuses, and the nucleolus disappears. During this period the number of chiasmata diminishes. By the end of diakinesis, in general the homologues are held together only at their ends, a process that has formerly received the name of *terminalization.*

Meilotic Division I

In *prometaphase* I condensation of the chromosomes reaches its maximum. The nuclear envelope breaks down, and the spindle microtubules become attached to the centromeres. Each homologue is attached to one of the poles by the homologue is attached to one of the poles by the homologue centromere, and the two sister sister chromatids behave as a functional unit.

In *metaphase* I the chromosomes become arranged at the equator. If the bivalent is long, it presents a series of annular apertures between the chiasmata in perpendicularly alternating planes. If the chromosomes are short, they have a singe annular aperture. Meiotic metaphase I can be distinguished under the microscope (from that occurring in mitosis or in metaphase II) by the fact that the homologues are still attached by chiasmata at their ends while the centromeres are being pulled towards the poles.

In *anaphase* I the sister chromatids of each homologues, united by their centromeres, move towards their respective poles. The short chromosomes, generally connected by a terminal and unterminalized chiasmata, is delayed. In side view, anaphase chromosomes show different shapes, depending on the position of the centromere. It should be recalled that by recombination, segments were transposed between two of the chromatids of each homologue. Thus when the homologous paternal and maternal chromosomes separate in anaphase, their composition is different from that the originals. Two of their chromatids are mixed; he other two maintain their initial nature with reference to a single locus.

Telophase. I begins when the anaphase groups arrive at their respective poles. Chromosomes may persist for some time in a condensed state showing all their morphologic characteristics. Following telophase is a short *interphase* which has characteristics similar to mitotic interphase. Sometimes interphase may persist for a considerable length of time.

The result of the first meiotic division is the formation of the daughter nuclei, which in animals are called spermatocytes II (in the male) and oocyte II plus the first polar body (in the female). At the interphase between the two meiotic divisions *three is no replication of the chromosomes.* These are now haploid in number although each one consists of two chromatids.

The short prophase II is followed by the formation of the spindle, which marks the beginning of metaphase II. At metaphase II chromosomes becomes arranged on the equatorial plane, and the sister centromers separate; the two sisterchromatids go towards the opposite poles during anaphase II. Since the longitudinal halves of each parental chromosomes (chromatids) separate in this division each of the four nuclei of telophase II has one chromatid, and each nucleus has a has a haploid number of chromosomes.

We have just described that in meiotic division I there is a separation of the homologous centromeres (and chromatids), while in meiotic division II, as in mitosis, separation of the sister centromeres (and chromatids) occurs. The differences observed are of great importance in explaining the general mechanism of chromosome distribution in mitosis and meiosis. On the basis of early observations by Schrader, in 1951 Ostergren postulated that the differences in chromosome distribution inn mitosis and meosis resulted from the different orientation of the kineytochores.

In fact, inn mitosis the sister chromatids have their kinetochores arranged back to back to back in meiotic division I the kinetochrores of the sister chromatids lie side by side. The factor that determines the distribution of the chromosomes is their initial interaction with the spindle just after disintegration of the nuclear envelope. This interaction occurs by way of spindle microtubules that run from the kinetochrore towards one to the poles and that determine the particular movement of the given chromosome.

The problem of chromosome distribution has been studied in experiments using microsurgery in grasshopper spermatocytes. Meiocytes in the I and II divisions were fused by mechanical means to produce one cell containing the two spindles —one spindle with bivalents, the other with unpaired chromosomes. It was observed that in each spindle the chromosomes by haved as in the original cells. Furthermore it was possible to detach a bivalent from division II and to place it in the division I spindle. In this arrangement the attached bivalent also divided in the original fashion (i. e., the two sister chromatids moved together of each pole). in all the other chromosomes, single sister chromatids moved to each pole. These and other experiments suggest that inn the mechanism of chromosomal distribution the key role is played by the orientation of the kinetochore, which determines the nucleation and preferential polarization of microtubules.

Genetic consequences—Meiosis

We know that the essential processes of meiosis are:

- Pairing or synapsis,
- Recombination and chiasmata,
- Segregation of the homologous chromosomes.

After the description of the entire process of meiosis, the third point become apparent. Meiosis has not only resulted in the halving of the chromosome number but has also segregated each member (*i.e.*, chromatid) of the homologous pair into four different nucele. One point that is of special interest is that the homologous centromeres separate in anaphase I, and the sister centromeres, in anaphase II. but since crossing over mixes the homologous chromatids of the bivalents, both meiotic division are needed to segregate the genes contained in the chromatids. We shall now see that segregation of the homologues and *recombination* have important genetic consequences. From the genetic viewpoint, meiosis can be considered as a mechanism for distributing the genes between the gametes, permitting their recombination and reandom segregation. The chiasmata are followed through diplonema, metaphase I, and anaphase I, and then into anaphase II. It is obvious that each one of the four gametes has a different genetic constitution. As a result of the crossing over, each chromosome does not consist solely of maternal or paternal material but of alternating segments of each.

The random assortment of the genes is due not only to crossing over but also to the fact that the distribution of the chromosomes in the first and second division is normally random. To simplify the matter, the three paternal and maternal homologues are shown separating into different cells. Since this separation is a random process, the resulting cell will contain eight different chromosomal combinations, even in the absence of crossing over, In a human (with 23 pairs of chromosomes), the possible chromosomal combinations in gametes will be an immense number, 2^{22} or. 8, 388, 608. Meiosis is the mechanism by which genetic variation is brought about and knowledge of its mechanism is prerequisite for the understanding of the chromosomes basis of genetics. We mentioned earlier that intermediary or sporic meiosis is characteristic of flowering plants. Meiosis takes place at some time intermediate between fertilization and the formation of gametes.

In higher plants the reproductive organsanthers in male and ovary or pistil in female produce microspores and megaspores, respectively. The cells that undergo meiosis to produce megaspors are called megasprorocytes. Microspores are produced by microsporocytes (pollen mother cells). Each microsporocyte gives rise, by meiosis, to four functional microspores. Each megasporocyte produces four megaspores by meiosis of which three degenerate. The remaining megaspore develops into the female gametophyte, which gives rise to the egg cell.

In plants, microsporoses, and megaspore are not the final gametes. Before fertilization they not the final gametes. Before fertilization, they undergo two mitotic divisions in the anther or three in the ovary to produce the male and female gametophytes, respectively Fertilization in plants is a complex phenomenon.

Each pollen grain or *microspore* (haploid) carries. two sperm nuclei. One of them fertilizes the egg nucleus to give a *diploid zygote,* which will eventually from the embryo of new plant (*i.e., sporophyte*) the other sperm nucleus fuses with two polar nuclei form a *triploid endosperm nucleus,* which by mitotic divisions will give rise to the *endosperm,* which contains the nutritive material for the embryo. Thus, the seed is a mosaic of tissues consisting of the diploid zygote, the triploid endosperm, and the diploid integument, which are of maternal origin.

Meiosis Age in Human Female

The primary female germ cells (i e., *gonocytes*) appear in the human embryo in the wall of the yolk sac at about 20 days and during the fifth week migrate to the gonadal ridge. By mitosis, they form oögonia and become surrounded by the follicular cells. forming the *primary follicles.*At the end of the third month of prenatal development, oögonia enter meiosis, becoming oocytes I and are then arrested at the stage of diplonema until sexual maturity at about 12 years. (This prolonged diplonema stage is sometimes called *dictiotene.,*) During this long phase, the chromosomes rennin in a rather unconditioned state and resemble the *lamphrush chromosomes.*

The number of oocytes in newborn female has been estimated at about 1,000,000; however, most of these degenerate. By the age of 7 years, there are some 300, 000 oocytes, but only about 400 reach maturity between 12 and 50 years Thus, meiosis may last as long as 50 years! This may explain the increase in the incidence of chromosomal aberration with the increasing age of the mother.

When the ovum is released into the *oviduct, the* first meiotic division occurs, giving one polar body. It is only when the ovum is fertilized by the spermatozoon that the second meiotic division takes please giving the second polar body.

As result of oogenesis, only one viable egg is produced; the haploid nucleus is known as the *female pronucleus*. The polar bodies degenerate and do not take part in embryogenesis,

Meiosis in Human Male

In the human male the primitive gonad also starts with the migration of the primary gonocytes into the gonadal redges, and later these are into the *seminiferous tubules*. These contain spermatogonia that enter meiosis at puberty, with the beginning of sexual life. In contrast to oogenesis in the female, spermatogenesis continues in the male until advanced age. Completion of this process takes about 24 days and the meiotic prophase 1 occurs in 13 to 14 days. Four viable spermatozoa are formed from each meiotic cycle. The transformation of spermatids into future sperm cells occurs by a complex process called *spermiogenesis*, which can be studied in histology textbooks.

Fertilization Interaction Between the Gametes

Feritilization is a specific process in the sense that, as a rule, the sperm of one species can not fertilize the egg of another. The egg produces a protein called *fertilizin* which reacts with an anti-fertilizin on the surface of the sperm. This is a species-specific interaction by which the cells are attached to each other The sperm, together with one centriole, then penetrates the egg. The male nucleus swells and becomes the so-called *male pronucleus* and then fuses with the female pronucleus, thus beginning the first division of the egg.

Thus the sperm in addition to acting as a stimulus of cell division, contributes:

- One set of haploid chromosomes carrying the parternal genes,
- A centriole, which will function in cell division during cleavage.

BIOCHEMISTRY OF MEISIS

Studies carried out mainly in *Lillium* have permitted the identification of the distinctive metabolic characteristics of meiosis which are not found in mitosis. Recently similar studies have been extended to meicytes of male mammals using special fractionation techniques. From the biochemical point of view the following four stages of prophasic meosis have been specifically studied:

Prezygonema

We have already mentioned that the premiotic S phase is longer than in somatic cells. In fact, it may last 100 to 200 times longer; this is due primarily to the reduced frequency in the number of initiation points during DNA duplication Another important difference observed in microsporocytes of *Lillium* is that 0.3 to 0.4per cent of the DNA remains unreplicated.

Zygonema

During zygonema the above-mentioned unreplicated DNA is synthesized stage in coordination with chromosome pairing. This late-replicating Z-DNA appears to be essential to synapsis, since if it is inhibited, the cells are arrested at zygonema. This Z-DNA has a higher buoyant density than the rest of the DNA and it is rich in GC bases. It is thought that Z-DNA contains highly repeated nucleotide sequences which are distributed over all *Lillium* chromosomes and may function in chromosome synapsis.

Another biochemical characteristic of this period is the appearance of a DNA. *binding protein* which is associted with a lipoprotein fraction and which catalyzes the reassociation of single-stranded DNA. This protein has been called *reassociation or-protein,* and it has been suggested that it may play a role in the alignment of homologous stretches of DNA.

Pachynema

Protein synthesis seems essential to maintain chromosome pairing during pachynema. If protein synthesis is inhibited at late zygonema the chromosome pairs fall apart and are not reconstituted. During pachynema, there is also a DNA synthesis which is of a smaller magnitude than that of zygonema. This. P-DNA synthesis is apparently related to the process of recombination. A molecular model based on a synthesis of repair-DNA that takes place when the two homologous chromatids recombine. It is thought that at the end of zygonema, an *endonuclease* is activated that produces nicks in the two DNA molecules that are aligned for recombination. After recombination gaps and overlaps of short nucleotide sequences can occur. The overlaps are excised by exonuclease, and the gaps are filled in by DNA polymerase and DNA ligase. The repair mechanism probably explains the small amount of DNA synthesis observed at pachynema.

9

Characteristics of Cell Differentiation

Cell differentiation is the process by which stable differences arise between cells. All higher organisms develop from a single cell, the fertilized ovum, which gives rise to the various tissues and organs. The question of how an apparently structureless egg converts itself into a complex and highly organized embryo has interested scientists since the time of Aristotle 2000 years ago, and this still remains one of the major unanswered questions of biology.

In most animal species females produce large unfertilized *eggs* which contain most of the materials and nutrients required to form an embryo. Development is triggered by *fertilization.* The sperm contributes a small, condensed nucleus (*male pronucleus*), which rapidly enlarges in the egg cytoplasm, fuses with the *female pronucleus,* and finally divides The fertilized egg then undergoes a series of very rapid cycles consisting of DNA synthesis followed by cell divisions. These divisions are called *cleavage, because* unlike normal cell division, the cytoplasm is partitioned without growth.

Then the cells form a hollow sphere (*blastula*) in which tissues are not yet evident. Some of the cells then invaginate in a series of cell movements known as *gastrulation,* and the first signs of morphologic differentiation appear. These very complex changes take place in a comparatively short time. For example, in the South African frog *Xenopus laevis,* a swimming tadpole containing most differentiated tissues (such as blood, nerve eye, muscle, and so forth) hatches only 72 houses after fertilization. In molecular terms, cell differentiation means *variable gene activity* in different cells of the same organism. Cell specialization involves the preferential synthesis of some specific proteins (*e.g.*, hemoglobin in erythrocytes, antibodies in plasma cells, and ovalbumin in oviduct). Each eukaryotic cell expresses only a small percentage of the genes it contains, and cells of different tissues express different sets of genes. However, it should be remembered that some genes are expected to be active in all types of cells, such as those genes required for building membranes, ribosomes, mitochondria, and the cytoskeleton, components which are essential to all cells. It is therefore clear that cell differentiation will be explained in molecular terms only when the complex mechanisms of gene regulation in higher cells are understood.

Nucleus and cytoplasm are interdependent and cell differentiation results from their interaction. We will analyse the three experimental systems which have contributed most to our knowledge in this field. (the) unicellular alga *Acetabularia,* frog (*Xenopus laevis*) oöcytes and eggs injected with somatic nuclei.

The level at which differential gene expression is controlled in early development and we will examine the possible role that is played in this process by "determinates," *i.e.*, molecules localized in specific regions of the egg cytoplasm. Whenever possible, we will refer to experiments work using *Xenopus laevis,* for it is in this animal that the biochemistry of early development is best understood.

BASIC FEATURES OF CELL DIFFERENTIATION

The Differentiated State is Stable and persistence

One of the principal characteristics of cell differentiation of higher cells is that once established, the differentiated state is very stable and can persist throughout many cell generations. For example, a neuron will persist as such throughout the lifetime of a individual, and a cell committed to become a skin cell will gradually keratinize and eventually die. These persistent changes are very different from the type of regulation involved in enzyme induction and repression in bacteria, which is specially designed to respond rapidly to changes in the environments. *E, coli* responds to the addition, of lactose to the culture medium by synthesizing B-galactosidase a few minutes later, and when lactose is exhausted, enzyme synthesis is stopped with equal speed.

Cell differentiation of induced in the organism by various stimuli, but once it has been established it can persist even in the absence of the initial., it can persist even in the absence of the initial stimulus.

This is demonstrated, for example, by the fact that differentiated cloned cell lines, such as steroid-secreting cell lines and neuroblastoma cell lines (which differentiate into neurons), are able to grown indefinitely in vitro. Without this type of data it could have been argued that differentiation persists in the organism because the stimulus that has given rise to it also persists. In virtro, although such a stimule cannot be present, the differentiated state is maintained over many cell generations.

Determination

In many cases, before it is possible to recognize morphologically that a cell has differentiated, there is a period during which the cell is already committed to a particular change. In this interval, called *determination,* the cell has already focused its capacity for differentiation on a specific pathway.

One of the classic examples of determination is provided by the *imaginal discs of Drosophila.* The discs are groups of cells which are present in the larva in an undifferentiated form, but which upon metamorphosis will give rise to legs, wings, antennae, and so forth. Each disc is predetermined to become a particular type of adult structure.

This commitment is very stable and heritable, as shown by the disc transplantation experiments carried out by Hadorn. Discs transplanted into the abdomen of adult files remained undifferentiated but could be stimulated to differentiate by transplantation into a larva where the hormonal conditions would induce metamorphosis. Discs transplanted serially in the abdomens of adult flies for 9 years (1800 cell generations) were still able, after metamorphosis, to differentiate into structures belonging to the original disc.

Inter dependency of Nucleus and cytoplasm

Nucleus and cytoplasm are interdependent; one cannot survive without the other. The cytoplasm provides most of the energy for the cell through oxidative phosphorylation (in mitochondria) and anaerobic glycolysis, and the cytoplasmic ribosomes contain most of the "machinery" for protein synthesis. On the other hand, the nucleus provides templates for specific synthesis (mRNA) and also supplies the other important RNA molecules (rRNA) and tRNA).

Any discussion of nucleocytoplasmic interrelations must consider:

- The mechanisms by which the genes contained in the chromosomes exert their control on the metabolic processes of the cytoplasm
- The mechanisms by which the cytoplasm influences gene activity.

Nuclear control of the cytoplasm was first considered during the last century by Balbiani in the so-called *merotomy* experiments.

Such enucleated fragments are able to sustain most cellular activities: *e.g.*, they can form a cellulose membrane and carry on photosynthesis (plant cells)), react to stimuli and ingest food (amoebae), activate cilia (ciliated cells), undergo cytoplasmic streaming, and so forth. However, these cells generally survive only a short time and are incapable of growth and reproduction.

Within 5 to 10 minutes after being enucleated by micromanipulation an ameba loses its surface tension, becomes spheroid, and develops numerous blunt pseudopodia. Its movements are slowed down and it can no longer digest foods. In this state it may survive for about 20 days.

If a nucleus is successfully implanted within three days, the ameba becomes extended, starts to move normally, and digests foods. Finally, it may even divide and eventually produce a culture mass.

ACETABULARIA

The unicellular marine alga *Acetabularia* has been invaluable for the analysis of nucleocytoplasmic interactions. This giant cell has a stem between 3 and 5 cm long and a cap 1 cm in diameter. The unique nucleus is located in the basal or *rhizoid* end of the cell. The great experimental advantage of *Acetabularia* is that the nucleus can be removed simply by cutting off the rhizoid. The resulting enucleated cells are active in photosynthesis and can survive for many weeks.

Each *Acetabularia* species has a particular cap morphology. If a nucleus from *A. crenulata* is implanted into an enucleated *a mediterranea,* the resulting cell will eventually develop a cap of the *A. crenulata* type. If two nucleate cells of different species are grafted together, a hybrid having a cap with intermediate morphology is formed. This clearly shows that the shape of the cap is determined by the nucleus.

Although dependent on the nucleus, the cytoplasm has some degree of functional autonomy. The information required to form the cap passes to the cytoplasm before cap formation, where it can be stored in an inactive form. *A. crenulata* does not form a cap in the darkness, only an elongated stem. If such a stem is nucleated and then exposed to light, a cap will develop. Therefore, the cap formation triggered by light is not dependent on new nuclear activity, and call the necessary information is present in the cytoplasm but is not converted to the final active form until a light stimulus is present.

The information for cap production is very stable; enucleated stalks maintained in the dark for weeks are still able to produce a cap upon illumination. This experimental result is probably due to the presence of stable mRNA are known to occur in other eukaryotic cells, for example, in the mammalian reticulocyte, which continues to produce hemoglobin for several days after the nucleus is eliminated in the course of normal erythropoiesis. The nucleus itself is dependent on the cytoplasm. *Acetabularia* has a giant nucleus in its rhizoid, but its volume decreases markedly when cytoplasmic energy production is impaired by darkness or by amputation of a large part of the cytoplasm.

Fusion of Cell

The fusion of cells by using inactivated *Sendai virus* (a member of the para-influenza viruses) provides the opportunity of placing a nucleus in a different cytoplasmic environment. Cell fusion can also be induced with other agents that affect membrane structure, such as polyethyleneglycol and lysolecithin. When different cells are fused, initially a *heterokaryon* is formed (*i.e.*, a single cell containing nuclei of two types). Eventually, both nuclei might enter mitosis synchronously, form a single metaphase plate, divide and produce a *hybrid cell line* (also known as a *synkaryon*). The cells of cell line have a single nucleus containing chromosomes of both parental nuclei.

In 1965 Henry Harris found that chick erythrocyte nuclei are reactivated when fused to HeLa cells (an undifferentiated human cell line so named because it was derived from the uterine carcinoma cells of a patient named *Henrietta Lacks*). These heterokaryons are of interest because the erythrocyte nucleus does not synthesize RNA or DNA.

Chick erythrocytes are terminally differentiated cells which have a very condensed nucleus and which are destined to die (Mammalian erythrocytes normally eliminate their nuclei during red blood cell maturation, while the red cells of birds, amphibians, and reptiles retain the nucleus, which becomes inactivated).

When fused to HeLa cells, the chick erythrocyte nucleus increases 20 times in volume, disperses its chromatin, resumes RNA synthesis, develops a nucleolus, and eventually replicates its DNA. This reactivation process is accompanied by the uptake of large amounts of human nuclear proteins which are thought to reactivate the erythrocyte nucleus this was shown by experiments in which heterokaryons were stained by immunofluorescence with antisera obtained from patients suffering from *lupus erythematosus.* (Lupus is a human autoimmune disease characterized by the production of antibodies against one's own nuclear proteins.) After fusion with HeLa cells, the chick nuclei stained intensely with the human nuclear antibodies.

From the From these experiments it is that the synthesis of macromolecules in a nucleus is controlled by the cytoplasmic environment. Even though erythrocytes are terminally differentiated cells, they can reassume RNA and DNA synthesis. A similar result is obtained with macrophage-HoLa cell heterokaryons. Rabbit macrophases do not synthesize DNA but will do so if fused to HeLa cells[7].

Only the cell cytoplasm is required to reactivate the chick erythrocyte nucleus. This was established by fusing erythrocytes to enucleated HeLa cells. Populations of enucleated cultured cells can be obtained with the use of *cytochalasin* B, a drug that affects predominantly actin microfilaments. Cells exposed to this drug cease moving and extrude their nuclei, which remain attached to the rest of the cell by only a cytoplasmic stalk.

Cultured cells that have been grown attached to a plastic surface are centrifuged in the presence of cytochalasin B, and the nuclei form a pellet, while the cytoplasm of the cells remains attached. The detached nudeus is surrounded by the cell membrane and a small amount of cytoplasm and is called a *karyoplast.* The enucleated cytoplasm is called a *cytoplasm.*

Enucleated cells are viable for at least two days after enucleation and perform many cell function is normally. For example, cell movements, pinocytosis, and contact inhibition are unaffected, and if cells are detached from the plastic surface by trypsinization, they can reattach and spread on a new Petri dish. Clearly, many cytoplasmic functions are independent of the cell nucleus.

Since karyoplasts are still surrounded by the cell membrane it is possible to fuse then to a different cytoplasm by using Sendai virus. *Reconstituted cells,* which arise from the fusion of a karyoplast and a cytoplast, can survive longer than enucleated cytoplasm, can synthesize RNA, and in some cases can undergo cell division.

Karyoplasts and cytoplasts from L cells (a mouse cell line) were prepared by centrifugation in the presence of cytochalasin B and were fused with Sendia virus, To distinguish true reconstituted cells from cells that failed to enucleate (karyoplast preparations are always contaminated by small numbers of intact cells). The uncleus was labeled with H-thymidine, and the cytoplasm, with latex beads of various diameters, (Cultured cells incubared in the presence of latex spares incorporate the beads by phagocytosis, thus providing a very useful cytoplasmic marker.)

Cytochalasin enucleation and cell reconstitution are very powerful techniques because they provide the opportunity of placing a nucleus (containing very little of its own cytoplasm) in an entirely new cytoplasmic environment.

In a few cases it has been possible to activate latent genes coding for specific proteins by fusion to differentiated cells. For example, albumin-secreting rat hepatoma cells fused with mouse lymphocytes (which do not produce albumin) will occasionally give rise to clones of hybride cells (synkaryons) which are able to produce both rat and mouse albumin. Somatic cell hybridization experiments shown that the patterns of nucleic acid synthesis and gene expression by a nucleus can be modified by substances present in the cell cytoplasm. Similar conclusions can be drawn from work with nuclei transplanted into *Xenopus laevis* oöcytes.

Cell Fusion Provide Pure Antibodies of Medical Importance

When animals are injected with a macromolecule of a shape that is recognized ad foreign to that individual (*e.g.*, a protein or complex polysaccharide), *antibodies* appear in the serum several days later. Antibodies are protein molecules that are able to bind specifically to the foreign *antigen*. Many different antibodies appear in the serum of an immunized animal, each one recognizing a different part of the antigen's shape (or different antigens if more than one was injected). Furthermore, individuals of the same species have different immunological responses,, so that two antisera directed against the same an tigen can in fact be very different. This is a major problem in medicine because, for example, success of failure of an organ transplantation depends on whether the donor and the recipient patient have the correct *histocompatibility antigens* on the cell surface. Clearly, antibodies that can be used throughout the world as standardized diagnostic reagents are highly desirable. This has now been made possible by using cell fusion techniques to create cell lines that produce only a single type of antibody, which can then be obtained in unlimited amounts.

Antibodies are produced by the lymphocytes. (When producing large amounts of antibody, lymphocytes adopt a special morphology and are also called *plasma cells.) Each antibody-producing cell can synthesize only one type of antibody.* Lymphocytes do not multiply in culture, but C. Milstein has developed a technique where by a *single* antibody-producing cell can be propagated indefinitely in culture by hybridization with a tumor cell.

Initially a mouse is immunized against the desired antigen. By using. Sendai virus, the lymphocytes from the spleen of this mouse are then fused to a mouse cell line derived from plasma cell tumors (called *myelomas* because they invade the bone marrow). The myeloma cell carries a mutation in the enzyme *hypoxanthing-guanine phosphorolysis transferase* and is therefore enable to grow in HAT medium. However, the resulting hybrid cells are able to grow in HAT medium and the clones that secrete the desired antibody have even identified and grown.

The fusion with the tumor cell "immortalizes" the spleen lymphocyle, which can then be grown indefinitely in culture or injected into mice, where the cells produce secreting tumour that can be maintained by serial transplantation. Since all of the cells of one clone are derived from a single lymphocyte, a *monoclonal antibody* of high purity is produced.

Monoclonal antibodies are currently used as anti-viral antibodies (*e.g.*, for diagnosis of influenza virus type and for the treatment of rabies), as diagnostic reagents in clinical biochemistry, and for the classification of human histocompatibility antigens. Monoclonal antibodies provide a good example of the fact that the practical applications of science cannot be predicted beforehand. Cell fusion was initially developed to study the seemingly specialized cell biology problem of nucleocytoplasmic interactions but eventually produced immunological tools of practical importance to all mankind.

Xenopus Oöcytes

Oöcytes used for experiments in differentiation are growing egg cells originally located in be ovary of the female frogs. Each oocyte requires three months or more to grow into a very large cell 1:2 mm in diameter. During this period the oocyte accumulates proteins for use during early embryogenesis, and its nucleus (also called the *germinal vesicle)* is very active in RNA synthesis. This high transcriptional activity can be visualized morphologically in the form of lampbrush chromosomes that have lateral loops packed with nascent RNA molecules.

During this growth period, the oocyte nucleus is arrested at the diplotene stage of meiotic prophase and therefore does not synthesize DNA, The stimulus to continue meiosis is provided by gonadotrophic hormones released from the frog's pituitary gland. In the laboratory this stimulus can be mimicked by the injection of mammalian hormones (or urine from a pregnant woman, thus providing a test that was much used for the detection of pregnancy). Hormone treatment rapidly induces maturation; the oöcytes are released from the ovary, progess through the first meiotic division while traversing the oviduct, and 15 hours later are laid as *eggs,* which are at meiotic metaphase II and ready to start development upon fertilization.

If nuclei are transplanted into oöcytes, the pattern of nucleic acid synthesis closely resembles that of the host cell. Brain nuclei (which synthesize RNA but not DNA) were injected into oöcytes at different stages of their life cycle. If injected into growing oöcytes (which synthesize RNA but not DNA). The brain nuclei incorporate H-uridine but not ^{3}H-thymidine in radioautographic studies. When nuclei are injected into maturing oöcytes isolated from the oviduct after hormone stimulation (such oöcytes are undergoing meiotic division (such oöcytes are undergoing meiotic division I and therefore have condensed chromosomes), the chromosome become condensed and mitotic spindles are formed. If not synthesize RNA but actively synthesize DNA after penetration of sperm or the micro-injection needle), they do not synthesize RNA, but will start replicating their DNA

The oöcyte cytoplasm not only affects the pattern of macromolecular synthesis but is also able to *reprogram* the expression of individual genes in transplanted nuclei. *Xenopus* oöcytes that were injected with a suspension of about 200 HeLa nuclei. The nuclei survive for several weeks inside the oöcytes and in that period synthesize substantial amounts of RNA.

In the first few days after injection the somatic nuclei enlarge to about ten times their original volume. The injected material can also be introduced into the oocyte nucleus in this case the somatic nuclei swell to about 100 times their initial volume. These changes are accompanied by chromatin dispersion and a massive uptake of proteins from the surrouding cytoplasm. The injected nuclei resemble the oocyte's own nucleus morphologically, and occasionally the human chromosomes form structures reminiscent of lampbrush chromosomes

The resemblance is not only morphological. The oocyte cytoplasm reprograms the gene expression of the injected nuclei in such a way that only those genes which are normally active in oöcytes are expressed. This can be determined by taking advantage of the fact that the RNAs synthesized by the injected nuclei in the course of several days accumulate in the cytoplasm, where they coder for new proteins.

De Robertis and Gurdon injected *Xenopus* cultured kidney cell nuclei into oöcytes of a different am phibian species (the salamander *Pleurodeles,* whose proteins can be readily distinguished from those of *Xenopus* by two-dimensional electrophoresis). Those genes that are normally expressed in kidney cells but not in oöcytes became inactive after injection into *Pleurodeles* oöcytes. More importantly, some oocyte-active genes (which were not expressed by the kidney cell nuclei) were activated by the oocyte cytoplasm. It is therefore clear that the oocyte contains components which can determine that a particular spectrum of protein-coding genes will be active and that others will be inative. The work with cell fusion and *Xenopus* oöcytes indicates that the cytoplasm of all cells contains components which determine the state of activity of nuclear genes. We will see that cytoplasmic gene-controlling substances (or "determinants"), which sometimes becomes unequally distributed in the cytoplasm, play a crucial role in establishing the initial steps of cell differentiation in early embryos and adult tissues.

CELL DIFFERENTIATION AND ITS MOLECULAR MACHANISMS

We will now discuss that at which level cell differentiation is controlled and we will examine the experimental evidence indicating that *the genome remains constant throughout cell differentiation*. Speciallized cells from different tissues contain the same kind and number of genes, and therefore the differences between specialized cells must be explained in terms of variable rates of gene expression. We will arrive at the conclusion the cell differentiation is most probably controlled at the transcriptional level.

Nuclear Transplantation

Much attention was devoted by nineteenth century embryologists to the possibility that cell differentiation could be due to the loss of those gens which are not expressed. However, massive loss of chromosomal material cannot be the cause of cell differentiation, since we know that the DNA content of diploid cells is constant in all tissues of an organism.

Rigorous proof that during embryonic differentiation there is no loss of genetic information came from the classical nuclear transplantation experiments performed by J.B. Gurdon; *Xenopus laevis* unfertilized eggs can be irradiated with ultraviolet light to destroy the endogenous nucleus and can then be injected with single *Xenopus* diploid nucleus. Nuclei obtained from *Xenopus* tadpole intestinal cell (which are clearly differentiated cells, since they have a "brush border" of microvilli) are able to sustain development of normal adult frogs which re fertile.

This demonstrates that the intestine cells, retained all the genes required for the complete life cycle of a frog. In addition since many forge can be obtained from the intestine of the same tadpole, nuclear transplantation allows a *clone* of genetically identical twins to be created. Development up to the stage at which the tadpole swims has been obtained using a variety of adult tissues, such as keratinizing skin cells and lymphocytes, thus demonstrating that the genes required to make nerve, blood, muscle, cartilage, and other tadpole tissues were not irreversibly inactivated in the donor nuclei.

Similar conclusions have been obtained using plant cells. It is common knowledge that whole plants can be grown from cutting. In some cases, such as in carrots, a complete plant can be grown from a single cultured cell.

Cell Differentiation can not be Explain by Gene Amplification

One way of obtaining differential gene activity would be to increase the number of copies of a given gene. There are tow circumstances in which *gene amplification* is known to occur. One is ribosomal DNA amplification observed in the oöcytes of amphibians and insects. *Xenopus* oöcytes selectively replicate their rDNA genes; a mature oocyte has 2 million copies of them (compared to 900 for a diploid somatic nucleus) and 1000 nucleoli in order to produce the vast amount of ribosomes (10^{12}) contained in a single egg.

The other case of amplification occurs in the DNA replication observed in certain puffs of *Rhyncosciara* salivary gland polytene chromosomes. However, *in both cases the specially amplified DNA is not passed on to future cell generations.* The salivary gland cells die at metamorphosis, and thee amplified oocyte DNA is not inherited by the embryo.

Since some differentiated cells produce large amounts of certain gene products (90 per cent of the protein synthesized in a reticulocyte is globin), it seemed possible that similar phenomena could occur in other genes. However, nucleic acid hybridization experiments have shown that there is only one copy haploid genome of globin or chicken ovalbumin genes in all tissues, regardless of whether the gene is preferentially expressed or not. In other words, a single globin gene, when fully activated, can give rise to all the globin required by a red blood cell. Gene amplification does not seem to be widespread phenomenon.

We have seen that differential gene expression does not result from changes in the *kind* or *number* of genes. Another possibility is that it could arise from changes in the *location* of genes in the chromosomes. Gene rearrangements and translocations occur in the case of antibody-producing cells. However, the DNA sequences adjacent to globin and ovalbumin genes have been analysed by cleaving cell DNA with restriction enzymes and were found to be the same regardless of the tissue from which the NDA was extracted. Similarly, no differences were found in the restriction enzyme fragments of *Drosophila* DNA extracted from embryonic and adult tissues.

Translational Control cannot explains cell Differentiation

Since the *genome remains constant throughout development,* the differences between cells must be due to differential gene activity; *i.e.,* some genes must be expressed at higher rate than others. This could be explained by differential *transcription* (RNA synthesis or processing) or by differential *translation* (protein synthesis) of genes. One passibility is that cells may have mechanisms by which some mRNAs are translated in a given cell type but not in others. Experiments in which mRNA is injected possibility.

Xenopus oöcytes, due to their large size (1.2 mm) and great hardiness (they can be injected with one twentieth of their volume without adverse effects) are very suitable for microinjection. A great number of mRNAs of animal and plant origin are efficiently translated in oöcytes (such as those coding for immunoglobulins, thyrogloblin, interferon, collagen, tobacco mosaic virus coat proteins and many others). This shows that oöcytes do not have a mechanism that excludes the translation of certain mRNAs.

Globin mRNA in oöcytes is very stable and can be continually translated for weeks; translation is extremely efficient, and each mRNA molecule can give rise to up to 100,000 molecules of globin proten. In fact living oöcytes translate mRNA much more efficiently than any cell-free system and have therefore becwidely used as a system for assaying mRNA is interesting to note both that purified mRNA requires great care when handled in viro avoid degradation by ribonucleases, which at very active and ubiquitions enzymes and the *homogenates of Xenopus* eggs and oöcytes have a great deal of RNA se activity. However, injected *living* oöcytes do not degrade mRNA is therefore clear that living cells can some time provide very favourable conditions fo.. biochemical analysis.

Although oöcytes are able to translate many different mRNAs, it could still be argued that this is because an oocyte is a non-specialized cell and that a specialized cell might not translate heterologous mRNAs. This proposition was tested in the elegant experiment. Rabbit globin mRNA was injected into fertilized eggs, which were then allowed to develop to the tadpole stage.

The tadpoles were dissected into an axial fragment (which contained almost exclusively muscle, nerve, and notochord) and the rest of the tadpole (which contained many tissues including blood-forming cells).

After labelling with amino-acetic, both fragments were analysed for synthesis of rabbit globin. Since the axial fragment (which does not contain blood-forming tissues) was found to synthesize globin to the same extent as the rest of the tadpole, it can be concluded that differentiated cells (muscle, nerve and notochord) can also translate heterologurs mRNAs.

The experiments thus far described suggest that cells cannot choose which *kinds* of specific mRNAs they will translate. However, cells can regulate the *amount* of proteins they synthesize. One example of this *quantitative* regulation is seen in sea urchin eggs, which undergo a ten-fold increase in protein synthesis after fertilization. This increase is due to the translation of mRNAs which were already present in the unfertilized egg, since the increase also occur if new RNA synthesis is eliminated (using actinomycin D or enucleation).

Although the *amount* of protein synthesized before and after fertilization are the same. Thus, protein synthesis is subject to *quantitative* control, but does not seem to be regulated in the *qualitative* way required to explain the expression of hemoglobin in red blood cells but not in other tissues.

Cell Differentiation is morelikely Controlled at the Level of Transcription

Several lines of evidence suggest that eukaryotic gene expression is regulated at the level of RNA synthesis. The clearest example is provided by polytene chromosomes in which transcription of genes can visualized in the from of puffs. Specialized cells have distinct patterns of puffing (and therefore of transcription) which vary between different tissues and respond to the hormone *ecdysone.*

Molecular hybridization studies have shown that only a fraction of the total genes are expressed in any one cell (only five to ten per cent of the single-copy genes). Furthermore, different sets of genes are transcribed in each specialized cell type. Although many RNAs are different, many others are shared by different cell types. The shared sequences represent the so-called *housekeeping genes* required for the survival of all cells, such as, for example, genes coding for somal and mitochondrial proteins, and so forth. The unshared sequences represent the so-called *luxury functions* which are characteristic of specialized cells, for example, hemoglobin in red blood cells ovalbumin in oviduct, keratin in skin.

Some of these proteins are expressed in very large amounts and it is possible to prepare the corresponding mRNAs can then be used to prepare *radioactive probes* for hybridization. (Highly labeled copies of the mRNA can be prepared with the enzyme reverse *transcriptase.* Since the enzyme requires a primer, oligo dT is first hybridized to the poly. A tail of the mRNA, and then a *complementry*-H^3 DNA copy or cDNA is obtained.) The ^{3}H-cDNA probe can then be used to measure the concentration of mRAN for that gene in various types of cells. For exmple ovalbumin sequences are not detectable in brain RNA. When hen oviduct is stimulated to secrete ovalbumin by the injection of estrogen, the increase in protein synthesis is paralleled by an increases in the specific mRNA. For all differentiated genes which have been examined, the experimental evidence is consistent with the view that *production of specialized proteins is due to differential gene transcription.*

Understanding the way in which eukaryotic transcription is regulated is of great importance in cell biology. Unfortunately, in eukaryotes we lack the sophisticated genetics which led to discovering how the *lac* operon works in *E. coli.* One way in which progress might be made is provided by *Xenopus* oöcytes microinjected with cloned eukaryotic DNAs. DNA injected into *Xenopus* oöcytes is transcribed only when introduced into the oocyte nucleus DNA injected into the cytoplasm is not transcribed and is rapidly degraded.

DNA injected into the nucleus seems to be transcribed accurately and associate with histones to form chromatin, and in some cases the mRNAs synthesized are translated into the correct protein products.

The synthesis of an SV 40 protein in oöcytes injected with DNA from SV40 virus. It is possible that living oöcytes injected with various purified eukaryotic DNAs (or with somatic nuclei) might provide a good assay system for factors (such as proteins or RNAs) presumed to be of importans or RNAs) presumed to be of importance in controlling chromatin transcription.

Cytoplasmic Determinants

We will now examine how the first differences appear between cells of the early embryo. As we have seen in the first part of this chapter, the cytoplasm of oöcytes and other cells contains molecules that influence the activity of the nucleus. Eggs have in their cytoplasm such subbstances, and in some cases they are localized in specific regions of the egg. As development proceeds, these substances, which are called *diterminants* of development, are unequally distributed into the cytoplasm of certain groups of cells which become committed to a particular type of cell differentiation.

Eggs are in general very large cells which stockpile many of the molecules required for early development. Fore example, a *Xenopus* egg contains about 100,000 times more RNA polymerases, histones, mitochondria, and ribosomes than a normal adult *Xenopus* somatic cell. One reason for accumulating these ready-made materials rather than making them *de novo* during early embryogenesis is the extraordinarily rapid rate of cell division during *cleavage*. At the mid-blastula stage, *Xenopus* replicates its DNA every 10 minutes, a process that takes 19 hours in the somatic cells of adults. Similarly, *Drosophila* blastulae double their DNA content in 3.4 minutes. As shown for both amphibia and *Drosophila.* this very short S phase is due to an increase in the number of replication forks.

This rapid pace allows little time for new RNA and protein synthesis. but it during this period that the first differences between nuclei are established. At least in some cases these difference result from the presence of *determinants* synthesized during oogenesis and localized in some regions of the egg cytopolasm.

The best example of *determinats* in development is provided by the *germ plasm.*Amphibian eggs contain in their vegetal (yolky) pole a specialized region of cytoplasm which can be recognized morphologically by the presence of special granules. *Drosophila* eggs also have an equivalent region located in the posterior poleof the egg (which is therefore called the *pole-plasm*). This cytoplasm has the property of inducing germ cell formation, *i.e.,* those cells that contain the germ plasm will eventually become the germ cells of the new organism.

When the posterior poles of eggs are irradiated with ultraviolet light, sterile (but otherwise normal) animals are obtained. However, if UV-treated egges are injected with pole-plasm of normal egges, fertile flies are obtained. Furthermore, if cytoplasm containing the germ cell determinants is injected into the anterior part of a *Drosophila* egg, germ cells develop in an anterior position.

We Known that determinats are in fact important in establishing early differences between cells, they cannot entirely explain development. Fore example, there is no evidence of cytoplasmic localization in mammalian eggs. Furthermore. as development advances, *cell interactions* become increasingly important.

At gastrulation extensive cell movements and migrations occur, and different types of cells interact with each other in the phenomenon known as *embryonic induction.* Notochord tissue induces the overlying ectodermal to become eye lens. These inductions are mediated by diffusible substances, but in spite of numerous attempts to isolate them, their chemical nature remains unknown.

It is possible that the same principles involved in the action of egg determinants could also apply to *adult cells.* In the initial part of this chapter we saw that all cells contain in their cytoplasm molecules that can reprogram nuclear gene expression. Cells continually exchange information between nucleus and cytoplasm so that gene prudently accumulated in the cytoplasm, upon cell division they can become unequally distributed between daughter cells giving rise to two different cell types.

This could occur in adult and embryonic cells. In differentiation of adult tissues it is frequently observed that only one of the daughter cells becomes specialized; the other one remains as *stem cell.* Which is able to divide again. This occurs in red blood cell differentiation and could occur in skin and intestinal epithelium, in with the dividing cells are located in certain regions of the epithelium (attached to the basal membrane or at the bottom of the intestinal crypts).

The hypothetical mechanism is supported by experimental evidence, During nerve cell differentiation in grasshoppers, some cell divisionresult in the formation of a neuron (ganglion cell) and a stem cell (neuroblast) which are always in the same position and morphologically recognizable. By introducing a needle at mitosis, it is possible to rotate the spite of this the resulting daughter cells still have the neuron and stem cell in the normal position.

This shown that the ability to become a neuron does not depend on particular chromosome set but rather on the type of cytoplasm inherited by the daughter cell.

Intimate Ration of Cell Deafferentation with cellular with Cellular Interactions

Cell interactions are intimately related to the biology of cell differentiation. As development proceeds, cell movement and interaction become increasingly important. Fore example, as gastrulation, cell migrations produce an invagination of the embryo, and eventually some of the invaginated cells (notochord) induce the ectodermal to become nerve tissue. In multicellular organism the cells of the different tissues must cooperate in a harmonious way.

Cells can interact with each other by way of long-range interactions. Interactions over longer distances are mediated by the diffusion of chemical substances as in the case of the embryonic induction discussed earlier. In many cases the interaction depends on short-range action or *cell contact.* The importance of the recognition properties of the cell surface can be realised immediately by thinking of the development of the nervous system, in which millions of neurons must find their specific partners to establish the synoptic junctions that give rise to neuronal circuits of immense complexity.

Despite many attempts, the chemical nature of the diffusible substances involved in embryonic induction remains unknown. Long-range interactions are better understood in the case of the cellular slime mold *Dictyostelium discoideum.* This organism lives in forests, under layers of decomposing leaves. While food is plentiful (*Dictyostelium* feeds on soil bacteria), the organism lives as unicellular amoebae which multiply by mitosis.

When food becomes scarce, some amoeba start secreting a diffusible substance that attracts more amebae. A *nucleation centre* is formed into which streams of amebae converge, until finally a *slug* or *pseudoplasmodium* is formed. The slug contains millions of cells and is able to migrate for considerable distances (towards light, where the spores will have a better chance to disperse). Eventually the slug differentiates, producing a stalk on top of which spores devlop. By sporulating, *Dictyostelium* ensures survival during periods of low food supply.

The *Dictyostelium* life cycle has two distinct stages, one of unicellular life and a multicellular stage in which millions of cells move and differentiate in a co-ordinated way. This whole process is triggered by a substance that attracts the cells together. The attractant has been identified and is cyclic AMP. Many cells use cyclic AMP as an intracellular singnal to regulate metabolism. Cellular slime moulds use cyclic AMP as an *extracellular messenger* to communicate positional information to distant cells.

10

Cell Cycle Regulation

In 1858 the pathologist Rudolph Virchow coined the cell doctrine which states that "When a cell arises, there must have been a previous cell, just as animals can only arise from animals and plants from plants." This doctrine is founded on the understanding that whether one is examining a single-celled organism or an animal as complex as man, the product is a result of repeated rounds of cell growth and division.

Most eukaryotic cells will proceed through an ordered series of events in which the cell duplicates its contents and then divides into two cells. This cycle of duplication and division is called the cell cycle. In order to maintain the fidelity of the developing organism this process of cell division in multicellular organisms must be highly ordered and tightly regulated. The loss of control will lead to abnormal development and is the cause of cancer. The eukaryotic cell cycle is composed of four steps, or phases, as depicted in the figure below.

Of the four phases depicted in the Figure, the two critical steps are DNA replication, which occurs during S-phase, and the physical process of cell division which occurs during M-phase (for mitosis). If we start at the beginning of the process, a cell undergoes a period where all of the necessary machinery for the process of DNA replication is synthesized. This process occurs during what is referred to as a gap between S-phase and M-phase and is termed G1. Following DNA replication, the cell pauses in another gap phase termed G2 where all the machinery necessary for cell division is synthesized. M-phase is composed of two discreet steps: mitosis, which constitutes the pairing and separation of the duplicated chromosomes, and cytokinesis which is the physical process whereby the cell splits into two daughter cells.

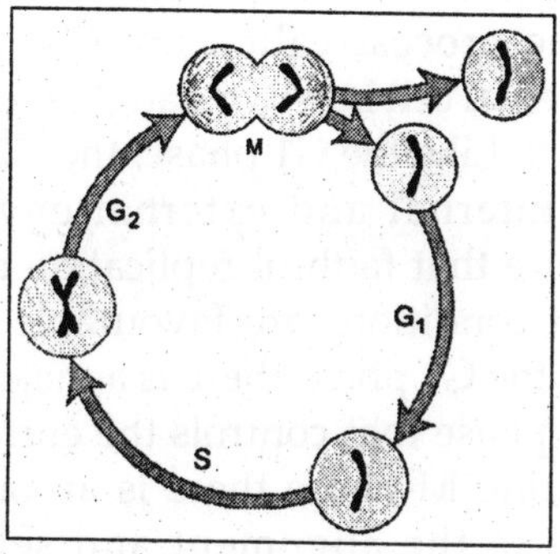

Fig. Phases of a Typical Eukaryotic Cell Cycle

Not all cells continue to divide during the life-span of an organism. Many cells undergo what is referred to as terminal differentiation and become quiescent and no longer divide. Cells in this phase of their life-cycle are said to reside in another gap phase called G0.

Under certain conditions, such as that resulting from an external signal stimulating cell growth, cells can exit the quiescent state and re-enter the cell cycle.

THE ORDERED STEPS IN THE CELL CYCLE

G1: The first gap in the normal cell cycle is called G1 and is the period when the necessary proteins for DNA replication are synthesized. However, this phase of the cell cycle is not only characterized by synthesis of replication machinery. During this period the cell must monitor both the internal and external environments to ensure that all the preparations for DNA synthesis have been completed and that overall conditions for cell division are favourable. As discussed below, there is a major check-point in a normal cell cycle that is critical for ensuring that all is well for the cell to enter S-phase.

S-phase: The duplication of the cellular content of DNA occurs during S-phase, so-called because this is the phase when DNA is synthesized. This phase of the cell cycle is the longest taking 10-12 hours of a typical 24hr eukaryotic cell cycle.

G2: During the second gap phase of the cell cycle the cell undertakes the synthesis of the proteins required to assemble the machinery required for separation of the duplicated chromosomes (the process called mitosis) and ultimately division of the parental cell into two daughter cells (the process termed cytokinesis). Like the G1 phase, the G2 phase is also a stage when the internal and external environments are monitored to ensure that faithful replication of the DNA has occurred and that conditions are favourable for cytokinesis. In addition, as for the G1 phase there is a major check-point at the end of the G2 phase that controls the entry into M-phase.

M-phase: During M-phase there is an ordered series of events that leads to the alignment and separation of the duplicated chromosomes (called sister chromatids) This process is divided into distinct steps that were originally identified and characterized through light microscopic observations of dividing cells. The steps of mitosis are termed prophase, prometaphase, metaphase, anaphase and telophase. Although cytokinesis is the process by which the parental cell is physically separated into two new daughter cells, it actually begins during anaphase. The processes that occur during M-phase require much less time than those of S-phase, generally lasting only 1-2 hrs.

During prophase the duplicated chromosomes condense while outside the nucleus the mitotic spindle assembles between the two centrosomes. The centrosome is an organelle that serves as the main microtubule organizing centre that is involved in the attachment of microtubules to the sister chromatids. During prometaphase the nuclear membrane breaks apart and the chromosomes can attach to spindle microtubules and begin active movement. During metaphase the chromosomes are aligned at the equator of the spindle midway between the spindle poles. The sister chromatids are attached to opposite poles of the spindle.

During anaphase the sister chromatids synchronously separate to form the two sets of daughter chromosomes. Each sister chromatid is slowly pulled towards the spindle pole it faces. During telophase the two daughter chromosomes arrive at the spindle poles and decondense. A new nuclear envelope forms around each set of chromosomes which forms the two new nuclei. This process marks the end of mitosis and sets the stage for cytokinesis.

CHECKPOINTS AND CELL CYCLE REGULATION

It should seem obvious that the processes that drive a cell through the cell cycle must be highly regulated so as to ensure that the resultant daughter cells are viable and each contains the complement of DNA found in the original parental cell. There are many "parts" to the systems that control the transit through a eukaryotic cell cycle.

These "parts" include mechanisms to control the timing of events so that each individual process is turned on and off at the appropriate time, mechanisms to initiate each event in the correct order and to also ensure that each event is triggered only once per cell cycle, controls to ensure events occur in a linear, irreversible direction, redundancy, or back-ups to ensure the cycle functions properly even in the context of some malfunctioning parts, and systems that are adaptable so that cell cycle events can be modified in the context of different cell types and/or environmental conditions. Many of the most important discoveries about the mechanisms that control events of the cell cycle were elucidated using yeasts which are single cell eukaryotes. By analysis of various mutants that inactivated genes encoding essential components of cell cycle control systems in yeast many important control genes were identified. These genes were identified as cell division cycle genes or cdc genes. Thus, many cell cycle control genes in mammalian cells are also called cdc genes. Much of the control of the progression through the phases of a cell cycle are exerted at checkpoints. There are many such checkpoints but the two most critical are those that occur near the end of G1 prior to S-phase entry and those near the end of G2 prior to mitosis.

As indicated above, there is the need for cell cycle control mechanisms to exert their influences at specific times during each transit through a cell cycle. The heart of this timing control is the responsibility of a family of protein kinases that are called cyclin-dependent kinases, CDKs. The kinase activity of these enzymes rises and falls as the cell progresses through a cell cycle. Different CDKs operate at different points in the cell cycle. As would be expected, the oscillating changes in the activity of CDKs leads to oscillating changes in the phosphorylation of various intracellular proteins.

These phosphorylations alter the activity of the modified proteins which then effect changes in events of the cell cycle. The cyclical activity of each CDK is controlled by a complex series of proteins, the most important of which are the cyclins, hence the name of the enzymes as cyclin-dependent kinases. The CDKs are absolutely dependent upon their interaction with the cyclins for activity. Unless they are tightly bound CDKs have no kinase activity. The cyclins were originally idenitified because they undergo a cycle of synthesis and degradation at specific points in each cell cycle. Thus, whereas the levels of the various CDKs remain fairly constant throughout the cell cycle, their activities changes in concert with the fluctuations of the cyclins. Four different classes of cyclins have been defined on the basis of the stage of the cell cycle in which they bind and activate CDKs. These four classes are G1-cyclins, G1/S-cyclins, S-cyclins, and M-cyclins.

The cyclin nomenclature and associated CDK in mammalian cells are listed here:

Cyclin-CDK Complex	Cyclin	CDK Partner
G1-CDK	cyclin D	CDK4, CDK6
G1/S-CDK	cyclin E	CDK2
S-CDK	cyclin A	CDK2
M-CDK	cyclin B	CDK1

- There are three D cyclins in mammals: D1, D2, and D3
- CDK1 is the same as CDC2 in fission yeast and CDC28 in budding yeast

The G1-cyclins are not found in all eukaryotic cells but in those where they are synthesized they promote passage through a restriction point in late G1 called Start. The G1/S-cyclins bind to their cognate CDKs at the end of G1 and it is this interaction that is required to commit the cell to the process of DNA replication in S-phase. The S-cyclins bind to their cognate CDKs during S-phase and it is this interaction that is required for the initiation of DNA synthesis. The M-cyclins bind to their cognate CDKs and in so doing promote the events of mitosis.

Although CDKs are inactive unless bound to a cyclin, there is more to the activation process than just the interaction of the two parts of the complex. When cyclins bind to CDKs they alter the conformation of the CDK resulting in exposure of a domain that is the site of phosphorylation by another kinase called CDK-activating kinase (CAK). Following phosphorylation the cyclin-CDK complex is fully active. In addition to control of CDK kinase activity by cyclin binding and CAK phosphorylation, control is exerted to inhibit CDK activity through interaction with inhibitory proteins as well as by inhibitory phosphorylation events.

Thus, there is extremely tight control on the overall activity of each CDK. One of the inhibitory kinases that phosphorylates CDKs is called Wee1. The inhibitory phosphorylations are removed through the action of a phosphatase called CDC25. The action of these two regulatory enzymes on CDK activity is most important at the level of the M-CDK activity at the onset of mitosis. Proteins that bind to and inhibit cyclin-CDK complexes are called CDK inhibitory proteins (CKI, for cyclin-kinase inhibitor).

Mammalian cells express two classes of CKI. These are called CIPs for CDK inhibitory proteins and INK4 for inhibitors of kinase 4. The CIPs bind and inhibit CDK1, CDK2, CDK4, and CDK6 complexes, whereas the INK4s bind and inhibit only the CDK4 and CDK6 complexes. There are at least three CIP proteins in mammalian cells and these are identified as $p^{21Cip1/WAF1}$ (gene symbol=CDKN1A), p^{27KIP1} (gene symbol=CDKN1B), and p^{57KIP2} (gene symbol=CDKN1C).

The expression of each of these CIPs is controlled by specific events that may have occurred during cell cycle transit. For example p^{21Cip1} expression is induced in response to DNA damage. This induction is under the control of the action of the tumor suppressor protein p53. There are at least four INK4 proteins that are each identified by their molecular weights: p^15INK4B^, p^16INK4A^, and p^18INK4C^ (these were the first 3 characterized) as well as p^19INK4^. The p^16INK4A^ protein is also a tumor suppressor since loss of its function leads to cancer. All the INK4 proteins contain 4 tandem repeats of a sequence of amino acids that were first identified in ankyrin and are thus referred to as ankyrin repeats.

As indicated above, many cells reside in a resting or quiescent state but can be stimulated by external signals to re-enter the cell cycle. These external growth promoting signals are the result of growth factors binding to their receptors. Most growth factors induce the expression of genes that are referred to as early and delayed-response genes. The activation of early response genes occurs in response to growth factor receptor-mediated signal transduction resulting in phosphorylation and activation of transcription factor proteins that are already present in the cell. Many of the induced early response genes are themselves transcription factors that in turn activate the expression of delayed-response genes. In the context of the cell cycle, these delayed-response genes encode proteins of the G1-CDK complexes.

One such early response gene is the proto-oncogene MYC. With respect to the cell cycle some of the genes turned on by activation of MYC are cyclin D, proteins of the ubiquitin ligase complex called SCF (Skp1/cullin/F-box protein) and the members of the E2F transcription factor family.

There are six members of the E2F family: E2F1 through E2F6). The synthesis of cyclin D will result in the activation of G1-CDK complexes. The synthesis of components of SCF leads to the degradation of p27KIP1 which normally inhibits G1-CDK complexes. The synthesis of E2F family members results in increased synthesis of proteins involved in DNA synthesis as well as the synthesis of the S-phase cyclins A and E and CDK2. Regulation of E2F activity by the tumor suppressor pRB will be discussed below. The cyclical degradation of the cyclins is effected through the action of several different ubiquitin ligase complexes. The action of ubiquitin ligases in protein turn-over is discussed in more detail in the Protein

Modifications page. There are two important ubiquitin ligase complexes that control the turn-over of cyclins and other cell cycle regulating proteins. One is the SCF complex which functions to control the transit from G1 to S-phase and the other is called anaphase promoting complex (APC) which controls the levels of the M-phase cyclins as well as other regulators of mitosis.

One important function of APC is to control the initiation of sister chromatid separation which begins at the metaphase-anaphase transition. The attachment of the sister chromatids to the opposite poles of the mitotic spindles occurs early during mitosis. The ability of the sister chromatids to be pulled apart is initially inhibited because they are bound together by a protein complex termed cohesin complex.

The cohesin complex is deposited along the chromosomes as they are duplicated during S-phase. Anaphase can only begin with the disruption of the cohesin complex. The breakdown of the cohesin complex is brought about as a consequence of the activation of the ubiquitin ligase activity of the APC. APC targets a protein called securin. Securin functions to inhibit the protease called separase and the action of separase is to degrade the proteins of the cohesin complex, thus allowing sister chromatid separation.

TUMOR SUPPRESSORS AND CELL CYCLE REGULATION

Tumor suppressors are so called because cancer ensues as a result of a loss of their normal function, *i.e.* these proteins suppress the ability of cancer to develop. It would seem obvious, therefore, that one import function of tumor suppressors would be control of the progression of a cell through a round of the cell cycle. If cells are able to synthesize damaged DNA before it is repaired or to divide when the DNA is damaged then the resulting daughter cells can pass on the resultant DNA damage to their progeny.

The result can be catastrophic resulting in cancer. For this reason, the two most important check points in the eukaryotic cell cycle are the G1-S transition and the entry into mitosis. The former prevents DNA replication prior to repair of damaged DNA and the latter prevents damage that may have occurred to the DNA during replication to propagated into daughter cells during mitosis. Following the isolation and characterization of two tumor suppressor genes in particular it was found that they function to control the ability of cells to progress through these two important checkpoints. The protein encoded by the retinoblastoma susceptibility gene (pRB) and the p53 protein are both tumor suppressors. The function of pRB is to act as a brake preventing cells from exiting G1 and that of p53 is to inhibit progression from S-phase to M-phase.

The best understood effect of G1-CDK activity is that exerted on transcription factors of the E2F family, hereafter referred to simply as E2F. In the context of the cell cycle regulation, E2F activates the expression of cyclin A, cyclin E and CDK2. These proteins are components of the S-CDK complexes necessary for progression through S-phase. The activity of E2F is itself controlled via interaction with pRB. When pRB binds E2F it can no longer function as a transcription factor as it is sequestered in the cytosol.

Interaction of pRB and E2Fcorrelates to the state of phosphorylation of pRB and the affinity between the two proteins is highest when pRB is hypophosphorylated. Phosphorylation of pRB is maximal at the start of S phase and lowest after mitosis and entry into G1. Stimulation of quiescent cells with mitogen induces phosphorylation of pRB, while in contrast, differentiation induces hypophosphorylation of pRB. One of the most significant substrates for phosphorylation by the G1 cyclin-CDK complexes is pRB. When pRB is phosphorylated by G1 cyclin-CDK complexes it releases E2F allowing E2F to transcriptionally activate its target genes.

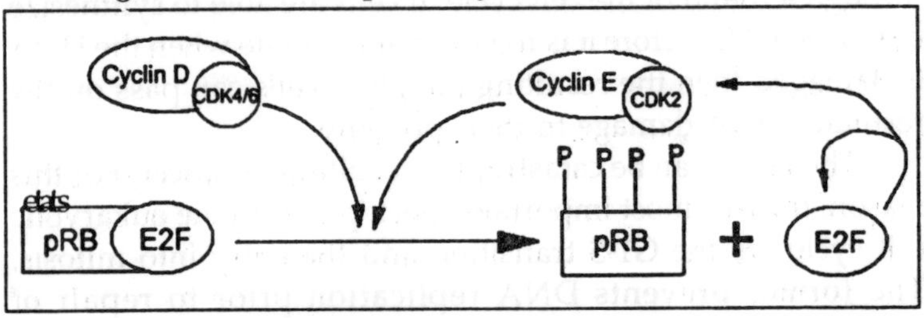

Fig. Regulation of E2F by pRB

When E2F activates the expression of S-CDK complex proteins these complexes also target pRB for phosphorylation, thus maintaining the cell in a pro cell cycle progression. One major function of the p53 protein, which is active as a homotetrameric transcription factor, is to serve as a component of the checkpoint that controls whether cells enter as well as progress through S-phase. The action of p53 is induced in response to DNA damage. Under normal circumstances p53 levels remain very low due to its interaction with a member of the ubiquitin ligase family called MDM2.

MDM2 is so named since it was isolated as an amplified gene in the tumorigenic mouse cell line 3T3DM. In response to DNA damage, *e.g.* as a result of uv-irradiation or irradiation, cells activate several kinases including checkpoint kinase 2 (CHK2) and ataxia telangiectasia mutated (ATM). One target of these kinases is p53. ATM also phosphorylates MDM2. When p53 is phosphorylated it is released from MDM2 and can carry out its transcriptional activation functions. One target of p53 is the cyclin inhibitor p21Cip1 gene.

Activation of p21Cip1 leads to increased inhibition of the cyclin D1-CDK4 and cyclin E-CDK2 complexes thereby halting progression through the cell cycle either prior to S-phase entry or during S-phase. As a consequence of p53-induced synthesis of p21 expression, there is a convergence between the roles of p53 and pRB in regulation of cyclin-CDK complexes. In either case the aim is to allow the cell to repair its damaged DNA prior to replication or mitosis. Even given the limited discussion of the functions of pRB and p53 it is still easy to understand how loss of either protein function can lead to aberrant cell cycle progression and the potential for the development of cancer.

THE MECHANICS OF CELL DIVISION

The process of cell division must occur in a highly ordered and accurate manner so as to ensure that each daughter cell receives an identical copy of the parental cells genome. The processes described above relate to the regulation of the progressive steps taken as a cell progresses to cytokinesis. This section will discuss the biochemical processes undertaken to effect accurate separation of the duplicated DNA (the sister chromatids) and cytokinesis.

Following duplication of the chromosomes the sister chromatids are held together through multisubunit protein complexes referred to as cohesins. These complexes are found all along the length of each chromatid as the DNA is replicated. Following DNA replication the chromosomes condense and this is the role of proteins called condensins. Chromosome condensation is the first easily identifiable sign that a cell is about to enter M-phase. Cohesins and condensins are structurally related and act in concert to prepare the chromosomes for mitosis.

The task of separating the sister chromatids such that each daughter cell receives one copy of each chromosome is carried out by the mitotic spindle which is composed of microtubules and several proteins that interact with them. An additional cytoskeletal structure is required for the actual separation of the cell into two new cells. This structure is referred to as the contractile ring.

The process of mitosis occurs through a highly ordered series of five steps as outlined above. The actual separation of the parental cell into two daughter cells can be considered the sixth step in mitosis. Whereas, prophase, prometaphase, metaphase, anaphase, and telophase occur in a strictly controlled sequential fashion, cytokinesis begins in anaphase and continues until the cell divides. During interphase the microtubule machinery is in a constant state of dynamic instability. Individual microtubules are growing or shrinking at any given moment.

During prophase the activation of M-CDK complexes initiates a change in the microtubule structures to one where there are a large number of shorter microtubules surrounding each centrosome. The centrosomes are cytoplasmic nucleation sites for the mitotic spindles. M-CDK complexes initiate these changes via the phosphorylation of microtubule motor proteins and microtubule-associated proteins (MAPs). During prometaphase the nuclear envelope abruptly breaks down as a consequence of M-CDK complexes phosphorylating the nuclear lamina. The dissolution of the nuclear membrane allows the microtubules to access the mitotic spindles.

When microtubules attach to the mitotic spindle they become stabilized. The microtubules eventually become attached at the kinetochore which is a complex protein structure that assembles onto the highly condensed DNA at the centromere. The chromosomes are pulled back and forth by the microtubules eventually becoming aligned equidistant from the two spindle poles. The alignment of the chromosomes forms the metaphase plate.

The chromosomes oscillate about the metaphase plate awaiting the signal that will induce the sister chromatids to separate. This phase of mitosis is referred to as the spindle-attachment checkpoint and it ensures that cells do not enter anaphase until all of the chromosomes are attached to both poles of the mitotic spindles. As described above, sister chromatids begin to separate with the activation of the APC. As each sister chromatid is pulled towards one of the poles of the mitotic spindle the kinetochore microtubules depolymerize.

By the end of anaphase, the daughter chromosomes have separated to opposite ends of the cells and have begun to decondense which signals the onset of telophase. Telophase is denoted by the reassembly of the nuclear envelope around each group of daughter chromosomes. During this process the lamins that make of the nuclear lamina are dephosphorylated allowing them to re-associate with the nuclear envelope. Following the formation of the new nuclear envelopes, the chromosomes decondense into their interphase state and transcriptional activity begins anew. The cell is now ready for the final process, complete separation into two daughter cells.

11

Properties of Stem Cells

Stem cells have the remarkable potential to develop into many different cell types in the body during early life and growth. In addition, in many tissues they serve as a sort of internal repair system, dividing essentially without limit to replenish other cells as long as the person or animal is still alive. When a stem cell divides, each new cell has the potential either to remain a stem cell or become another type of cell with a more specialized function, such as a muscle cell, a red blood cell, or a brain cell.

Stem cells are distinguished from other cell types by two important characteristics. First, they are unspecialized cells capable of renewing themselves through cell division, sometimes after long periods of inactivity. Second, under certain physiologic or experimental conditions, they can be induced to become tissue- or organ-specific cells with special functions. In some organs, such as the gut and bone marrow, stem cells regularly divide to repair and replace worn out or damaged tissues. In other organs, however, such as the pancreas and the heart, stem cells only divide under special conditions.

Until recently, scientists primarily worked with two kinds of stem cells from animals and humans: embryonic stem cells and non-embryonic "somatic" or "adult" stem cells. The functions and characteristics of these cells will be explained in this document. Scientists discovered ways to derive embryonic stem cells from early mouse embryos nearly 30 years ago, in 1981. The detailed study of the biology of mouse stem cells led to the discovery, in 1998, of a method to derive stem cells from human embryos and grow the cells in the laboratory. These cells are called human embryonic stem cells. The embryos used in these studies were created for reproductive purposes through in vitro fertilization procedures. When they were no longer needed for that purpose, they were donated for research with the informed consent of the donor. In 2006, researchers made another breakthrough by identifying conditions that would allow some specialized adult cells to be "reprogrammed" genetically to assume a stem cell-like state.

This new type of stem cell, called induced pluripotent stem cells (iPSCs), will be discussed in a later section of this document. Stem cells are important for living organisms for many reasons. In the 3- to 5-day-old embryo, called a blastocyst, the inner cells give rise to the entire body of the organism, including all of the many specialized cell types and organs such as the heart, lung, skin, sperm, eggs and other tissues. In some adult tissues, such as bone marrow, muscle, and brain, discrete populations of adult stem cells generate replacements for cells that are lost through normal wear and tear, injury, or disease.

Given their unique regenerative abilities, stem cells offer new potentials for treating diseases such as diabetes, and heart disease. However, much work remains to be done in the laboratory and the clinic to understand how to use these cells for cell-based therapies to treat disease, which is also referred to as regenerative or reparative medicine. Laboratory studies of stem cells enable scientists to learn about the cells' essential properties and what makes them different from specialized cell types. Scientists are already using stem cells in the laboratory to screen new drugs and to develop model systems to study normal growth and identify the causes of birth defects.

FUNCTIONS OF STEM CELL

Stem cells differ from other kinds of cells in the body. All stem cells-regardless of their source-have three general properties: they are capable of dividing and renewing themselves for long periods; they are unspecialized; and they can give rise to specialized cell types. Stem cells are capable of dividing and renewing themselves for long periods. Unlike muscle cells, blood cells, or nerve cells-which do not normally replicate themselves-stem cells may replicate many times, or proliferate. A starting population of stem cells that proliferates for many months in the laboratory can yield millions of cells. If the resulting cells continue to be unspecialized, like the parent stem cells, the cells are said to be capable of long-term self-renewal. Scientists are trying to understand two fundamental properties of stem cells that relate to their long-term self-renewal:

- Why can embryonic stem cells proliferate for a year or more in the laboratory without differentiating, but most non-embryonic stem cells cannot; and
- What are the factors in living organisms that normally regulate stem cell proliferation and self-renewal?

Discovering the answers to these questions may make it possible to understand how cell proliferation is regulated during normal embryonic development or during the abnormal cell division that leads to cancer. Such information would also enable scientists to grow embryonic and non-embryonic stem cells more efficiently in the laboratory. The specific factors and conditions that allow stem cells to remain unspecialized are of great interest to scientists.

It has taken scientists many years of trial and error to learn to derive and maintain stem cells in the laboratory without them spontaneously differentiating into specific cell types. For example, it took two decades to learn how to grow human embryonic stem cells in the laboratory following the development of conditions for growing mouse stem cells.

Therefore, understanding the signals in a mature organism that cause a stem cell population to proliferate and remain unspecialized until the cells are needed. Such information is critical for scientists to be able to grow large numbers of unspecialized stem cells in the laboratory for further experimentation. Stem cells are unspecialized. One of the fundamental properties of a stem cell is that it does not have any tissue-specific structures that allow it to perform specialized functions.

For example, a stem cell cannot work with its neighbours to pump blood through the body (like a heart muscle cell), and it cannot carry oxygen molecules through the bloodstream (like a red blood cell). However, unspecialized stem cells can give rise to specialized cells, including heart muscle cells, blood cells, or nerve cells. Stem cells can give rise to specialized cells. When unspecialized stem cells give rise to specialized cells, the process is called differentiation. While differentiating, the cell usually goes through several stages, becoming more specialized at each step.

Scientists are just beginning to understand the signals inside and outside cells that trigger each stem of the differentiation process. The internal signals are controlled by a cell's genes, which are interspersed across long strands of DNA, and carry coded instructions for all cellular structures and functions. The external signals for cell differentiation include chemicals secreted by other cells, physical contact with neighbouring cells, and certain molecules in the microenvironment. The interaction of signals during differentiation causes the cell's DNA to acquire epigenetic marks that restrict DNA expression in the cell and can be passed on through cell division.

Many questions about stem cell differentiation remain. For example, are the internal and external signals for cell differentiation similar for all kinds of stem cells? Can specific sets of signals be identified that promote differentiation into specific cell types? Addressing these questions may lead scientists to find new ways to control stem cell differentiation in the laboratory, thereby growing cells or tissues that can be used for specific purposes such as cell-based therapies or drug screening. Adult stem cells typically generate the cell types of the tissue in which they reside. For example, a blood-forming adult stem cell in the bone marrow normally gives rise to the many types of blood cells.

It is generally accepted that a blood-forming cell in the bone marrow-which is called a hematopoietic stem cell-cannot give rise to the cells of a very different tissue, such as nerve cells in the brain. Experiments over the last several years have purported to show that stem cells from one tissue may give rise to cell types of a completely different tissue.

This remains an area of great debate within the research community. This controversy demonstrates the challenges of studying adult stem cells and suggests that additional research using adult stem cells is necessary to understand their full potential as future therapies.

EMBRYONIC STEM CELLS

Embryonic stem cells, as their name suggests, are derived from embryos. Most embryonic stem cells are derived from embryos that develop from eggs that have been fertilized in vitro-in an in vitro fertilization clinic-and then donated for research purposes with informed consent of the donors. They are not derived from eggs fertilized in a woman's body. The embryos from which human embryonic stem cells are derived are typically four or five days old and are a hollow microscopic ball of cells called the blastocyst. The blastocyst includes three structures: the trophoblast, which is the layer of cells that surrounds the blastocoel, a hollow cavity inside the blastocyst; and the inner cell mass, which is a group of cells at one end of the blastocoel that develop into the embryo proper.

Embryonic Stem Cells Grown in the Laboratory

Growing cells in the laboratory is known as cell culture. Human embryonic stem cells are isolated by transferring the inner cell mass into a plastic laboratory culture dish that contains a nutrient broth known as culture medium.

The cells divide and spread over the surface of the dish. The inner surface of the culture dish is typically coated with mouse embryonic skin cells that have been treated so they will not divide. This coating layer of cells is called a feeder layer. The mouse cells in the bottom of the culture dish provide the inner cell mass cells a sticky surface to which they can attach. Also, the feeder cells release nutrients into the culture medium. Researchers have devised ways to grow embryonic stem cells without mouse feeder cells. This is a significant scientific advance because of the risk that viruses or other macromolecules in the mouse cells may be transmitted to the human cells.

The process of generating an embryonic stem cell line is somewhat inefficient, so lines are not produced each time an inner cell mass is placed into a culture dish. However, if the plated inner cell mass cells survive, divide and multiply enough to crowd the dish, they are removed gently and plated into several fresh culture dishes. The process of re-plating or subculturing the cells is repeated many times and for many months. Each cycle of subculturing the cells is referred to as a passage. Once the cell line is established, the original cells yield millions of embryonic stem cells. Embryonic stem cells that have proliferated in cell culture for six or more months without differentiating, are pluripotent, and appear genetically normal are referred to as an embryonic stem cell line. At any stage in the process, batches of cells can be frozen and shipped to other laboratories for further culture and experimentation.

Tests used to Identify Embryonic Stem Cells

At various points during the process of generating embryonic stem cell lines, scientists test the cells to see whether they exhibit the fundamental properties that make them embryonic stem cells. This process is called characterization. Scientists who study human embryonic stem cells have not yet agreed on a standard battery of tests that measure the cells' fundamental properties.

However, laboratories that grow human embryonic stem cell lines use several kinds of tests, including:

- Growing and subculturing the stem cells for many months. This ensures that the cells are capable of long-term growth and self-renewal. Scientists inspect the cultures through a microscope to see that the cells look healthy and remain undifferentiated.
- Using specific techniques to determine the presence of transcription factors that are typically produced by undifferentiated cells.

- Using specific techniques to determine the presence of paricular cell surface markers that are typically produced by undifferentiated cells.
- Examining the chromosomes under a microscope. This is a method to assess whether the chromosomes are damaged or if the number of chromosomes has changed. It does not detect genetic mutations in the cells.
- Determining whether the cells can be re-grown, or subcultured, after freezing, thawing, and re-plating.
- Testing whether the human embryonic stem cells are pluripotent by 1) allowing the cells to differentiate spontaneously in cell culture; 2) manipulating the cells so they will differentiate to form cells characteristic of the three germ layers; or 3) injecting the cells into a mouse with a suppressed immune system to test for the formation of a benign tumor called a teratoma. Since the mouse's immune system is suppressed, the injected human stem cells are not rejected by the mouse immune system and scientists can observe growth and differentiation of the human stem cells. Teratomas typically contain a mixture of many differentiated or partly differentiated cell types-an indication that the embryonic stem cells are capable of differentiating into multiple cell types.

ADULT STEM CELLS

An adult stem cell is thought to be an undifferentiated cell, found among differentiated cells in a tissue or organ that can renew itself and can differentiate to yield some or all of the major specialized cell types of the tissue or organ.

The primary roles of adult stem cells in a living organism are to maintain and repair the tissue in which they are found. Scientists also use the term somatic stem cell instead of adult stem cell, where somatic refers to cells of the body (not the germ cells, sperm or eggs). Unlike embryonic stem cells, which are defined by their origin (the inner cell mass of the blastocyst), the origin of adult stem cells in some mature tissues is still under investigation. This finding has led researchers and clinicians to ask whether adult stem cells could be used for transplants. In fact, adult hematopoietic, or blood-forming, stem cells from bone marrow have been used in transplants for 40 years. Scientists now have evidence that stem cells exist in the brain and the heart. If the differentiation of adult stem cells can be controlled in the laboratory, these cells may become the basis of transplantation-based therapies.

The history of research on adult stem cells began about 50 years ago. In the 1950s, researchers discovered that the bone marrow contains at least two kinds of stem cells. One population, called hematopoietic stem cells, forms all the types of blood cells in the body. A second population, called bone marrow stromal stem cells (also called mesenchymal stem cells, or skeletal stem cells by some), were discovered a few years later. These non-hematopoietic stem cells make up a small proportion of the stromal cell population in the bone marrow, and can generate bone, cartilage, fat, cells that support the formation of blood, and fibrous connective tissue.

In the 1960s, scientists who were studying rats discovered two regions of the brain that contained dividing cells that ultimately become nerve cells. Despite these reports, most scientists believed that the adult brain could not generate new nerve cells. It was not until the 1990s that scientists agreed that the adult brain does contain stem cells that are able to generate the brain's three major cell types-astrocytes and oligodendrocytes, which are non-neuronal cells, and neurons, or nerve cells.

Function of Adult Stem Cells

Adult stem cells have been identified in many organs and tissues, including brain, bone marrow, peripheral blood, blood vessels, skeletal muscle, skin, teeth, heart, gut, liver, ovarian epithelium, and testis. They are thought to reside in a specific area of each tissue (called a "stem cell niche"). In many tissues, current evidence suggests that some types of stem cells are pericytes, cells that compose the outermost layer of small blood vessels. Stem cells may remain quiescent (non-dividing) for long periods of time until they are activated by a normal need for more cells to maintain tissues, or by disease or tissue injury.

Typically, there is a very small number of stem cells in each tissue, and once removed from the body, their capacity to divide is limited, making generation of large quantities of stem cells difficult. Scientists in many laboratories are trying to find better ways to grow large quantities of adult stem cells in cell culture and to manipulate them to generate specific cell types so they can be used to treat injury or disease. Some examples of potential treatments include regenerating bone using cells derived from bone marrow stroma, developing insulin-producing cells for type 1 diabetes, and repairing damaged heart muscle following a heart attack with cardiac muscle cells.

Identification of Adult Stem Cells

Scientists often use one or more of the following methods to identify adult stem cells:

- Label the cells in a living tissue with molecular markers and then determine the specialized cell types they generate;
- Remove the cells from a living animal, label them in cell culture, and transplant them back into another animal to determine whether the cells replace (or "repopulate") their tissue of origin.

Importantly, it must be demonstrated that a single adult stem cell can generate a line of genetically identical cells that then gives rise to all the appropriate differentiated cell types of the tissue. To confirm experimentally that a putative adult stem cell is indeed a stem cell, scientists tend to show either that the cell can give rise to these genetically identical cells in culture, and/or that a purified population of these candidate stem cells can repopulate or reform the tissue after transplant into an animal.

Differentiation of Adult Stem Cells

Normal differentiation pathways of adult stem cells. In a living animal, adult stem cells are available to divide, when needed, and can give rise to mature cell types that have characteristic shapes and specialized structures and functions of a particular tissue. The following are examples of differentiation pathways of adult stem cells that have been demonstrated in vitro or in vivo.

Hematopoietic stem cells give rise to all the types of blood cells: red blood cells, B lymphocytes, T lymphocytes, natural killer cells, neutrophils, basophils, eosinophils, monocytes, and macrophages. Mesenchymal stem cells give rise to a variety of cell types: bone cells (osteocytes), cartilage cells (chondrocytes), fat cells (adipocytes), and other kinds of connective tissue cells such as those in tendons.

Neural stem cells in the brain give rise to its three major cell types: nerve cells (neurons) and two categories of non-neuronal cells-astrocytes and oligodendrocytes. Epithelial stem cells in the lining of the digestive tract occur in deep crypts and give rise to several cell types: absorptive cells, goblet cells, paneth cells, and enteroendocrine cells.

Skin stem cells occur in the basal layer of the epidermis and at the base of hair follicles. The epidermal stem cells give rise to keratinocytes, which migrate to the surface of the skin and form a protective layer. The follicular stem cells can give rise to both the hair follicle and to the epidermis.

TRANSDIFFERENTIATION

A number of experiments have reported that certain adult stem cell types can differentiate into cell types seen in organs or tissues other than those expected from the cells' predicted lineage (*i.e.*, brain stem cells that differentiate into blood cells or blood-forming cells that differentiate into cardiac muscle cells, and so forth). This reported phenomenon is called transdifferentiation.

Although isolated instances of transdifferentiation have been observed in some vertebrate species, whether this phenomenon actually occurs in humans is under debate by the scientific community. Instead of transdifferentiation, the observed instances may involve fusion of a donor cell with a recipient cell. Another possibility is that transplanted stem cells are secreting factors that encourage the recipient's own stem cells to begin the repair process. Even when transdifferentiation has been detected, only a very small percentage of cells undergo the process. In a variation of transdifferentiation experiments, scientists have recently demonstrated that certain adult cell types can be "reprogrammed" into other cell types in vivo using a well-controlled process of genetic modification. This strategy may offer a way to reprogram available cells into other cell types that have been lost or damaged due to disease. For example, one recent experiment shows how pancreatic beta cells, the insulin-producing cells that are lost or damaged in diabetes, could possibly be created by reprogramming other pancreatic cells.

By "re-starting" expression of three critical beta-cell genes in differentiated adult pancreatic exocrine cells, researchers were able to create beta cell-like cells that can secrete insulin. The reprogrammed cells were similar to beta cells in appearance, size, and shape; expressed genes characteristic of beta cells; and were able to partially restore blood sugar regulation in mice whose own beta cells had been chemically destroyed. While not transdifferentiation by definition, this method for reprogramming adult cells may be used as a model for directly reprogramming other adult cell types.

In addition to reprogramming cells to become a specific cell type, it is now possible to reprogram adult somatic cells to become like embryonic stem cells (induced pluripotent stem cells, iPSCs) through the introduction of embryonic genes. Thus, a source of cells can be generated that are specific to the donor, thereby avoiding issues of histocompatibility, if such cells were to be used for tissue regeneration. However, like embryonic stem cells, determination of the methods by which iPSCs can be completely and reproducibly committed to appropriate cell lineages is still under investigation.

Differences between Embryonic and Adult Stem Cells

Human embryonic and adult stem cells each have advantages and disadvantages regarding potential use for cell-based regenerative therapies.

- One major difference between adult and embryonic stem cells is their different abilities in the number and type of differentiated cell types they can become.
- Embryonic stem cells can become all cell types of the body because they are pluripotent. Adult stem cells are thought to be limited to differentiating into different cell types of their tissue of origin.
- Embryonic stem cells can be grown relatively easily in culture. Adult stem cells are rare in mature tissues, so isolating these cells from an adult tissue is challenging, and methods to expand their numbers in cell culture have not yet been worked out. This is an important distinction, as large numbers of cells are needed for stem cell replacement therapies.
- Scientists believe that tissues derived from embryonic and adult stem cells may differ in the likelihood of being rejected after transplantation. We don't yet know whether tissues derived from embryonic stem cells would cause transplant rejection, since the first phase 1 clinical trial testing the safety of cells derived from hESCS has only recently been approved by the United States Food and Drug Administration (FDA).
- Adult stem cells, and tissues derived from them, are currently believed less likely to initiate rejection after transplantation. This is because a patient's own cells could be expanded in culture, coaxed into assuming a specific cell type (differentiation), and then reintroduced into the patient. The use of adult stem cells and tissues derived from the patient's own adult stem cells would mean that the cells are less likely to be rejected by the immune system. This represents a significant advantage, as immune rejection can be circumvented only by continuous administration of immunosuppressive drugs, and the drugs themselves may cause deleterious side effects.

PROPERTIES OF A STEM CELL

Stem cells are different from the other types of cells in the human body. Although they can be harvested from various sources, they all share some of the same properties. In fact, stem cells are essentially a repair system in the human body. They can divide in unlimited numbers to replenish other cells, with new cells remaining a stem cell or becoming a more specialised cell with a specific purpose.

Stem Cells can Divide and Renew Themselves

Stem cells have a special ability to divide and renew themselves for extended periods of time. Cells such as muscle or nerve don't normally replicate, but stem cells will repeatedly proliferate. This basically means that they produce copies of themselves many times over. In fact, an initial population of stem cells can produce millions of cells within mere months in a laboratory setting. It is when the produced cells remain unspecialised that the long-term self-renewal is apparent.

Understanding why stem cells divide and renew as they do can help scientists discover what happens during dysfunctional cell division that leads to cancer cells. It can also help them answer the question of how the cells' proliferation and differentiation is regulated. Embryonic stem cells appear to be able to proliferate for a year or even longer without ever developing into specialised cells, contrasting with adult cells. Researchers are questioning why this occurs as well as what aspects regulate stem cell renewal.

Stem Cells are Unspecialised

The unspecialised nature of stem cells is an important one. It means that stem cells lack the specific parts that allow them to perform specialised functions in the body. If you think about your heart muscle, it functions to pump blood through your body. Also consider a nerve cell that sends signals to other nerve cells and the rest of your body, allowing you to move. A stem call does not have a specialised function but it has the capacity to differentiate into a specialised cell that can carry out these functions.

The exact factors that allow stem cells to be unspecialised are still somewhat of a mystery to researchers. Research has allowed scientists to grow stem cells in the laboratory, but the precise signals that trigger a stem cell to proliferate are still largely unknown. Once this is determined, scientists will be able to grow stem cells in the laboratory with greater success.

Stem Cells can give Rise to Specialised Cells

The ability of stem cells to give rise to specialised cells is a crucial one. In this process of differentiation, unspecialised stem cells produce specialised cells. It is thought that a cell's genes regulate the internal signals that trigger this process. These genes carry the specific code, or instructions, for all the parts and functions of a cell. External signals are those outside of the cell, which include chemicals released from other cells, physical connections with nearby cells and various other molecules in the surrounding area. Currently, scientists are searching for similarities and differences between the signals from one stem cell to another. Ultimately, answers will allow researchers to find ways to control stem cell differentiation. With control, scientists can more easily grow cells and tissues for specific functions. Stem cells are clearly unique cells with special attributes. It is these special attributes that offer the vast potential to treat numerous diseases and save lives. With further research, scientists will hopefully be able to understand more about the intriguing properties of stem cells and can then find ways to manipulate these properties for therapeutic value.

BENEFITS OF STEM CELLS

With all the controversy surrounding stem cells you may have missed hearing about many of the benefits for the health and medical fields. You may not even be aware that stem cells already have many applications for treating disease. Their potential to treat even more diseases in the future means that scientists are working hard to learn about how stem cells function and how they can treat some of the more serious diseases affecting the world.

Stem Cells and Human Development

Stem cells have enormous potential in health and medical

research but to fully harness this potential, scientists are studying how stem cells transform, or differentiate, into the diverse range of specialised cells that make humans what they are today. Because diseases such as cancer or conditions such as birth defects are thought to occur because of problems in the differentiation process, an understanding of the development that happens in normal cells will help scientists treat the developmental errors that can occur.

Stem Cells and Cell-Based Therapies

Another potential application of stem cells is to form cells and tissues for medical therapies. Currently, it is donated organs and tissues that are substituted for damaged or dysfunctional ones. Sadly, the number of people awaiting a transplant is much higher than the number of available organs.

Transplant waiting lists are enormous and many people die awaiting transplants. Stem cells offer a viable source of replacement cells to treat diseases and can potentially reduce the morbidity and mortality for those awaiting transplants. Some of the areas that stem cells can benefit include:

- Parkinson's disease
- Type I diabetes
- Arthritis
- Burn victims
- Cardiovascular diseases

By directing stem cells to differentiate into specialised cell types, there is the exciting possibility to provide a renewable source of replacement cells for those suffering from diseases.

The potential to reverse diseases is also not a foreign one. For example, a patient who has suffered from a heart attack and sustained heart damage could have the damaged tissue replaced by healthy new muscle cells. The destruction of brain cells in conditions such as Parkinson's disease can hopefully be reversed with the replacement of new, healthy and functioning brain cells. Even more promising is the potential to address genetic defects that are present from birth by restoring function and health with the introduction of normal healthy cells that do not have these defects.

Burn Victims

Burn victims tend to endure an enormous amount of pain from their wounds as well as frustration from the challenges of healing. Instead of donor tissues being donated, stem cells could be used to produce new and healthy tissues.

This is essentially similar to therapies already being used, such as bone marrow transplants, where stem cells create different specialised blood cells. Scientists aim to locate and remove specific stem cells from a tissue and then trigger them to differentiate outside of the body before transplanting them back into the patient to replace damaged tissues. In burn victims, a very small piece of the skin can be progressively grown, allowing doctors to cover a burn that is often much larger than the original size of the skin piece.

Stem Cells and Drug Testing

Stem cells have an important benefit for the pharmaceutical field. New drugs can be tested on stem cells to assess their safety before testing drugs on animal and human models. For example, a cancer cell line could be created to test an anti-tumour drug. If the conditions can be perfectly replicated, testing drugs could provide very accurate results.

The current benefits of stem cell usage are already well documented and it is expected that continued research will pave the way for new treatments. For those suffering from serious diseases, stem cells offer hope for effective treatment or perhaps even a reversal of the disease. Time will confirm the full success of stem cell therapies and continued research should teach us more about using stem cells to treat debilitating medical conditions.

CLASSIFYING STEM CELLS

Multipotent Stem Cells

Stem cells remain an exciting area of research but the concept is often still a confusing one for the public. Complicating the complex topic of stem cells is the separation of stem cells into various categories. One such category is multipotent stem cells and it carries traits that separate it from totipotent and pluripotent stem cells. Multipotent stem cells have the same basic features of all stem cells. As with all stem cells multipotent stem cells are unspecialized cells that have the ability to: Self-renew for long periods of time and differentiate into specialized cells with specific functions

A multipotent stem cell can give rise to other types of cells but it is limited in its ability to differentiate. These other types of cells are also limited in numbers. Examples of multipotent stem cells include those in the brain that give rise to different neural cells and glia or haematopoietic cells, which can give rise to different blood cell types, but they can't create brain cells. Bone marrow also contains multipotent stem cells that give rise to all blood cell types but not other cells. Hence, adult stem cells are considered multipotent because their specialization potential is limited to one or more cell lines. However, a multipotent stem cell known as a mesenchymal stem cell can give rise to several cell types. This particular stem cell has been found to give rise to bone, muscle, cartilage, fat, and other similar tissues.

Multipotent stem cells are essentially committed to produce specific cell types. They vary from stem cells such as pluripotent ones which can give rise to almost any cell type, or totipotent ones which can give rise to any cell, including the potential to create a complete organism. Pluripotent stem cells actually undergo specialization into multipotent stem cells, and then multipotent stem cells give rise to cells with a specific purpose and function. Multipotent stem cells can be likened to a family-although they do give rise to different cells, the cells themselves are within a certain family and therefore, are closely related. Multipotent stem cells are found in the tissues of adult mammals. It is thought that they are in most body organs, where they replace diseased or aged cells. Thus, they function to replenish the body's cells throughout an individual's life.

Benefits of Multipotent Stem Cells Favourite

There are numerous benefits and uses to multipotent stem

cells. Since multipotent stem cells are derived from pluripotent stem cells, these stem cells have already partially differentiated and they continue specializing as they develop.

They haven't yet been identified in all adult tissues but new research is regularly indicating a discovery of multipotent adult stem cells in new body tissues. Multipotent adult stem cells are particularly useful in transplants. They can be isolated, albeit often with difficulty, from a person's tissues and then guided to develop into a certain type of cell, before being transplanted back into the same patient. This avoids the immunological challenges of pluripotent foetal stem cell usage, where a patient's immune system could potentially reject a 'foreign' tissue. Another benefit is that the ethical debate and controversy involved in extracting foetal stem cells is avoided, because neither foetal tissues nor an aborted embryo are necessary for treatments.

A key type of multipotent stem cell that research is currently focused on is a neural cell. Neural cells give rise to nerve cells, which don't have the same turnover rate as do other cell types such as blood. Neural cells have now been isolated from the adult brain and foetal brain tissues, which means that these cells may then be triggered to divide, or differentiate, into new nerve cells. The implications for treating brain and spinal cord injuries are enormous and could essentially provide a reversal of these conditions.

Challenges and Current Multipotent Stem Cell Research Favourite

One challenge that scientists must address is whether or not a multipotent cell could actually be triggered to specialize into a cell type different from that of the original tissue. Current research on multipotent stem cells, however, is now contradicting the belief that multipotent adult stem cells are restricted to giving rise to the cell types corresponding to their specific tissues. The stem cells appear to be able to go beyond the previously believed boundaries for producing specific cell types, but they do so infrequently and only under narrow conditions. To understand how this occurs, the issue of stem cell plasticity has to be considered. Stem cell plasticity is a term used to describe the phenomenon of adult stem cells from one tissue generating the specialized cells of another tissue. For a long time, it has been believed that adult multipotent stem cells are organ-specific, so they only give rise to cell types in their residing tissues. This possible finding of plasticity has enormous potential for stem cell therapy.

It means that if scientists can re-focus this differentiation, a blood stem cell could, for instance, be used as a replacement for heart or nervous system tissues. There are still, however, numerous challenges that must be overcome before this idea can be put into practice. For now, it is more feasible that multipotent stem cells will have practical use in those tissues from which they are isolated. Multipotent adult stem cells have also not been found in all of the body tissues, which means that use of pluripotent foetal stem cells is the only opportunity to obtain some types of tissue. The difficulty in isolating large quantities of multipotent cells from adult tissues is an additional challenge that must be overcome.

In addition, the current time required for a multipotent stem cell to mature in a laboratory setting is quite lengthy but still necessary to provide sufficient cells for the treatment. Many patients are in dire need of the transplanted cells and simply can't stay alive long enough to receive the treatment. If the cells are intended to treat a disease that is genetically based, then it's likely that the isolated cells will also carry this genetic error and are ultimately rendered useless.

It's clear that there are many challenges to overcome before multipotent cells can be safely and regularly used, but their benefits are numerous and hold promise for treating disease. Their avoidance of the ethical issues associated with pluripotent and totipotent foetal cells also means that approval for research and therapeutic use is likely to be a more timely process.

Pluripotent Stem Cells

The concept of stem cells can seem a complicated one and you may have seen foreign words such as 'pluripotent' written in magazines or discussed on television. Stem cells describe all of the cells that can give rise to the different cells found in tissues. There are however, different types of stem cells. One such type is a pluripotent stem cell. Pluripotent stem cells are often termed 'true' stem cells because they have the potential to differentiate into almost any cell in the body. This means that under the right circumstances, a stem cell that is isolated from an embryo can produce almost all of the cells in the body. Yet after this embryonic development stage is over, the stem cells no longer have this unlimited potential to develop into all cell types. Their pluripotency is thus lost and they can only become certain types of cells. To understand how a cell becomes pluripotent, it helps to consider the human body in the very early stages of development. After an egg is fertilised by a sperm, a single cell results. This cell-the fertilised egg that is totipotent-has the potential to create an entire organism. In the initial hours and days following fertilisation, this single totipotent cell divides into more totipotent cells that are exact copies of the original.

Approximately four days after fertilisation, the totipotent cells start to specialise and form a cluster of cells known as a blastocyst. The blastocyst has yet another smaller group of cells known as the inner cell mass and it is these inner pluripotent stem cells that will go on to create most of the cells and tissues in the human body. These pluripotent stem cells are therefore different than totipotent stem cells because they don't develop into a complete organism. As such, a pluripotent cell won't give rise to the placenta or other tissues that are vital for foetal development. It will still develop into the other specialised cell types in the human body, such as nerve or heart cells.

You may have also heard the term 'stem cell line.' Stem cells from embryos can be used to create these pluripotent stem cell 'lines,' which are grown in the laboratory or cultured from foetal tissue.

Types of Pluripotent Stem Cells Favourite

There are several key types of pluripotent stem cells:

- Embryonic stem cells are isolated from the inner cell

mass of the blastocyst. The embryos are excess ones produced from in vitro fertilisation, but the practice is still controversial because it does destroy the embryo, which could have been implanted to create a baby.

- Embryonic germ cells are taken from aborted foetuses and these pluripotent cells are derived from very early cells. These early cells are those that can become sperm and eggs.
- Embryonic carcinoma or cancer cells are isolated from a type of tumour that sometimes occurs in a foetus.

Induced Pluripotent Stem Cells Favourite

Induced pluripotent stem cells (iPSCs) are adult cells that have been genetically reprogrammed to an embryonic stem cell-like state by being forced to express genes and factors important for maintaining the defining properties of embryonic stem cells.

Although these cells meet the defining criteria for pluripotent stem cells, it is not known if iPSCs and embryonic stem cells differ in clinically significant ways. Mouse iPSCs were first reported in 2006, and human iPSCs were first reported in late 2007.

Mouse iPSCs demonstrate important characteristics of pluripotent stem cells, including expressing stem cell markers, forming tumors containing cells from all three germ layers, and being able to contribute to many different tissues when injected into mouse embryos at a very early stage in development. Human iPSCs also express stem cell markers and are capable of generating cells characteristic of all three germ layers.

Although additional research is needed, iPSCs are already useful tools for drug development and modeling of diseases, and scientists hope to use them in transplantation medicine. Viruses are currently used to introduce the reprogramming factors into adult cells, and this process must be carefully controlled and tested before the technique can lead to useful treatments for humans.

In animal studies, the virus used to introduce the stem cell factors sometimes causes cancers. Researchers are currently investigating non-viral delivery strategies. In any case, this breakthrough discovery has created a powerful new way to "de-differentiate" cells whose developmental fates had been previously assumed to be determined.

In addition, tissues derived from iPSCs will be a nearly identical match to the cell donor and thus probably avoid rejection by the immune system. The iPSC strategy creates pluripotent stem cells that, together with studies of other types of pluripotent stem cells, will help researchers learn how to reprogram cells to repair damaged tissues in the human body.

Benefits and Future of Pluripotent Stem Cells Favourite

Pluripotent stem cells provide a chance to obtain a renewable source of healthy cells and tissues to treat a wide array of diseases such as heart disease and diabetes. Burn victims and those who suffer from autoimmune diseases such as Parkinson's can all potentially benefit from the use of pluripotent stem cells.

Pluripotent stem cells have a vast potential for the treatment of disease, namely because they give rise to the majority of cell types in the human body. These include muscle, blood, heart and nerve cells. Another potential use for pluripotent stem cells involves the generation of cells and tissues for use in transplantation.

Pluripotent stem cells can evolve into specialised cells that ultimately can replace diseased cells and tissues. Drug research is another area that pluripotent stem cells may benefit. Animals are a commonly used model to assess the safety and use of drugs. Instead of initially testing drugs on animals, they can be evaluated through testing on cells grown from pluripotent stem cells. Those drugs that appear tolerated and safe can then progress to testing on animals and finally, humans.

Challenges For Pluripotent Stem Cells Favourite

Often, the discoveries with the greatest therapeutic benefits present the most difficult challenges; this is particularly true for pluripotent stem cells. Researchers are attempting to learn ways of controlling the development process of pluripotent stem cells into the many different cell types in the human body. Another current challenge is that the cells used in research are rejected from a person's body due to their immune system. More controversial is the fact that many scientists and members of the public have ethical issues with the use of pluripotent stem cells from human embryos or foetal tissues.

The positive uses of pluripotent stem cells are enormous but new research and ethical challenges must be taken into account before the public can reap the full benefits. For those who suffer from the many diseases that may be treated by pluripotent stem cells, additional knowledge and research will hopefully come sooner rather than later.

Totipotent Stem Cells

You may already be familiar with the basics of stem cells but perhaps you weren't aware that there are different types, or potencies of stem cells. Totipotent stem cells are one of the most important stem cells types because they have the potential to develop into any cell found in the human body. In human development, the egg cell in a woman and the sperm cell from a man fuse together to form a single cell called the zygote. The zygote divides numerous times and forms cells that are the precursors to the trillions of cells that will eventually constitute the human body.

It may sound rather complex and in a sense it is, when you consider that a simple cell has the potential to develop into a specialised one with a specific job in the body. It is simple, however, in that it's a clear sequence of events that begins with an egg cell and a sperm cell. The zygote from that fusion of an egg cell and a sperm cell then begins cell divisions that are capable of forming the entire human body. It is these cells that are totipotent, so called because their potential is 'total.' Totipotent stem cells are perhaps the most versatile of the stem cell types. As explained, a totipotent zygote cell is created when a single celled sperm and egg unite. This totipotent fertilised egg has the potential to give rise to virtually all human cells, such as nerve or heart. It is during the early cell divisions in embryonic development that more totipotent cells are produced. Within several days, these totipotent cells divide and create replicas, therefore producing more totipotent cells.

It is after approximately four days that the cells begin to specialise into pluripotent cells, which can go on to specialise further but can't ever produce an entire organism as totipotent cells can. Basically, the pluripotent stem cell can do everything the totipotent one can except for creating an entire organism. A fertilised egg has totipotency, or total potential for about four days. Days after fertilisation, however, the totipotent stem cell divides and then matures to cause more specialised stem cells called pluripotent stem cells. Like totipotent stem cells, pluripotent stem cells can self-renew and give rise to trillions of cells in the body.

Special Qualities of a Totipotent Stem Cell Favourite

The earlier stage foetal stem cells are totipotent because:

- They have the ability to become any cell type in a fully developed human.
- They have the ability to replicate in unlimited numbers without losing their total potency.

The ability of totipotent stem cells to differentiate into any cell in an organism including embryonic tissue is an important distinguishing quality. A human totipotent stem cell can thus develop into any cell in the body, including placental cells.

Benefits of Totipotent Stem Cells Favourite

All stem cells have the potential to develop into different cell types, but totipotent cells can develop into any cell type, which makes them ideal for cell and gene therapies as well as tissue engineering for transplants and replacement of diseased cells. This means that the therapeutic value of totipotent stem cells is enormous. By learning about the process of division, we can find out what goes wrong in disease states and then investigate ways to prevent diseased cell production and division.

A formidable challenge is for researchers to appease the ethical concerns around the harvesting of totipotent foetal cells. This is balanced by supporters who cite the huge range of benefits to those suffering from disease. For now, there doesn't appear to be an answer to satisfy all members of the public as well as scientists and government agencies. The promise of effective treatments or even a cure for many debilitating diseases is, however, reason enough to search for treatments with totipotent stem cells that everyone can live with.

Unipotent Stem Cells

Due to intense interest in totipotent and pluripotent stem cells, unipotent stem cells have not received the same attention and research. They do, however, have vast potential to treat health conditions. A unipotent stem cell refers to a cell that can differentiate along only one lineage. The word 'uni' itself is derived from the Latin word 'unus,' meaning one.

Found in adult tissues, a unipotent stem cell, in comparison with other types of stem cells, has the lowest differentiation potential. This means that the cell has the capacity to differentiate into only one type of cell or tissue, which is lower in potential compared to stem cells that give rise to a broad range of cell types. Although the unipotent adult stem cells in the body's tissues will only give rise to one cell type, they do still have the important property of self-renewal that is shared by all stem cells. Also, despite their differentiation potential being limited, unipotent cells still have vast therapeutic potential to treat injuries and diseases.

A unique property of stem cells is their ability to differentiate, which means they can form specialized cells. Another special property is their ability to proliferate, or divide repeatedly.

As mentioned, unipotent stem cells have a very limited ability to differentiate relative to other stem cells such as pluripotent, totipotent or multipotent cells. Their ability to self-renew, however, does make them a valuable candidate for therapeutic use in treating disease. They are thus able to generate healthy and viable cells for transplant purposes.

Development of Unipotent Cell Favourite

Unipotent stem cells arise from multipotent cells. A multipotent stem cell is one can develop into a limited number of tissue types and it arises from totipotent and pluripotent stem cells, which can give rise to almost any specialized cell in the body. A unipotent stem cell will start to differentiate and give rise to a specific stem cell. The tissue-specific cell will then provide functional and structural components to a body tissue or organ.

Benefits of Unipotent Stem Cellsfavourite

Skin cells, which are in the epithelium, are one of the most abundant types of unipotent stem cells. The epithelium is the outermost tissue layer, which in itself has a top layer of dead squamous epithelial cells. This is similar to the mucus membranes that line our mouths and other body cavities. By taking a portion of a patient's own undamaged skin stem cells, sheets of skin can be developed for transplanting over burned areas of the patient's body. This technique is an important one for burn victims, particularly considering the pain and disfigurement that many burn victims are forced to experience during and after healing.

Unipotent Stem Cell Limitationsfavourite

A major limitation of this technology, however, is that it is extremely time consuming. It can take several weeks to grow a sufficiently sized piece of skin. Skin is an initial barrier to disease so the pressure to promote healing and replace burned skin is high as well as very time sensitive. Improved research into the ideal conditions to encourage unipotent stem cell growth should hopefully ensure that the success of this form of therapy grows.

STEM CELL THERAPY

Stem cell therapy is a set of techniques that aim to replace cells damaged or destroyed by disease with healthy functioning ones. The techniques themselves are still relatively new but their applications and benefits are broad.

Some of the diseases they can help include:

- Cancer
- Parkinson's disease
- Diabetes
- Various injuries (*e.g.* spinal cord)
- Eye Diseases

Types of Stem Cell Therapy

There are a number of stem cell therapies that are currently being investigated or used to treat a range of diseases.

These are:

- Adult stem cell transplants using bone marrow stem cells.
- Adult stem cell transplants using peripheral stem cells.
- Stem cell transplants using umbilical cord blood.
- Therapeutic cloning.

Benefits of Current Stem Cell Therapy

Although a great deal of stem cell therapy is still experimental, researchers have found the treatments promising for many diseases. Bone marrow stem cells, for example, are used to replace blood cells in people suffering from leukaemia and other cancers. Burn victims are also benefiting from stem cell therapy, which allows for new skin cells to be grafted as a replacement for damaged ones. New healthy cells can also be grown to treat eye diseases as well as replace dysfunctional tissues. The ultimate goal of stem cell therapy is to replace unhealthy cells with healthy ones and to do so efficiently, allowing proper cell functioning in the human body. In addition, the potential uses for stem cell therapy continue to grow as we learn more about the process whereby a healthy cell becomes a diseased one.

Challenges of Stem Cell Therapy

A major difficulty with stem cell therapy is to identify stem cells within an actual tissue culture. Cultures contain many different cells and it is a challenge to identify specific cell types. When the stem cells are identified and then isolated from tissues, the appropriate solutions must be created to trigger these cells to develop into the desired cell types.

Finally, there are additional issues even when cells are identified, isolated and grown. The new cells require implantation in a person and they must then essentially learn how to effectively function alongside a person's own tissues.

For instance, if you imagine a cardiac cell being implanted, think about the fact that it may not beat with the same rhythm of a person's heart cells and is thus ineffective. A person's immune system may also recognise the transplanted cells as foreign bodies and this can trigger an immune reaction that results in rejection of the new cells. The potential of stem cell therapy to ease human suffering and dramatically affect disease has motivated scientists to research ways of enhancing current stem cell therapies and develop new ones. Stem cell therapy remains a new science but the results have thus far been impressive enough that scientists are eagerly studying ways to treat the many diseases that you or a loved one may suffer from one day.

Bone Marrow Stem Cell Harvest

A bone marrow stem cell transplant uses stem cells derived from bone marrow to provide a fresh and healthy source of new blood cells which in turn, allows for a patient to receive higher doses of chemotherapy to treat certain types of cancer such as leukaemia. This ultimately means that a person has a better chance of surviving cancer. The bone marrow stem cells may be allogeneic and therefore donated by a family member of stranger, or they may be autologous, which utilizes a patient's own stem cells.

Importance of Bone Marrow Favourite

Bone marrow is the soft tissue found in the centre of bones. It is here that new blood cells are formed; stem cells are considered the parent cells of these blood cells. Stem cells in the marrow have the vital task of creating a person's three blood cell types.

These are:

- Red blood cells: transport oxygen.
- White blood cells: fight disease.
- Platelets: aid in clotting after injury.

Bone marrow stem cells are found in bone marrow and in a person's blood. After stem cells multiply, they form immature blood cells, which are then subject to a collection of changes that allow them to develop into mature blood cells. Once mature, the blood cells migrate from the marrow and are introduced into the bloodstream, where they provide important functions in keeping the body alive and healthy.

Effects Of Chemotherapy On Bone Marrow Favourite

Because chemotherapy operates by destroying the fast growing cancer cells, higher levels are generally more effective for killing cancer cells. Similar to cancer cells, however, bone marrow cells grow quickly and are also very sensitive to chemotherapy treatment. The chemotherapy can eventually destroy the marrow completely, which then prevents new and healthy blood cells from developing.

Autologous Bone Marrow Stem Cell Harvest Favourite

In an autologous transplant, a patient's own stem cells are used. The problem here is that the patient's bone marrow and peripheral blood usually have a lot of cancer cells, so isolating stem cells can't generally occur until the patient is in remission.

A patient will usually receive some chemotherapy to reduce cancer cells before stem cells are collected. The harvested stem cells are also treated to ensure that no cancer cells remain. Higher doses of chemotherapy are then given, sometimes alongside complete body radiation, to confirm that no cancer remains. Stem cells are then transplanted back into the body via a rapid injection. Stem cells will eventually migrate to the bone marrow, where they latch onto other cells there and develop into the different blood cells.

Allogeneic Bone Marrow Stem Cell Harvest Favourite

In the allogeneic transplant, a suitable donor is used; this may be a family member or a person unrelated to the patient. Similar to the autologous bone marrow stem cell harvest, a patient receives high doses of chemotherapy and possibly also radiation therapy. This therapy will destroy the patient's marrow and immune system, so that the patient's immune system can't attack the transplanted cells that are received from the donor. Graft rejection is a very real concern for allogeneic transplants, where the recipient's immune system recognizes the transplanted cells as 'foreign' and launches an immune attack. Prior to the transplant, red blood cells in the stem cell sample are usually identified and discarded and particularly so if the donor's blood group is different from that of the patient.

Stem cells are then infused into the patient via an intravenous line over several hours. Stem cells travel to the patient's bone marrow where they develop and produce the blood cells necessary for blood functioning. Patients may also still be given drug therapy for some time to reduce the chances of immune rejection. Bone marrow stem cell harvests are clearly a life saving technique for those suffering from certain cancers such as leukaemia. They are one of the 'older' stem cell therapies and have been proven effective for decades now. There are, however, still issues of rejection that warrant further development and refinement of stem cell harvesting techniques. It is hoped that scientists will continue to focus on research to improve the odds of success for this important treatment.

Disappointments in Stem Cell Therapies

Although there have been many breakthrough studies into stem cell therapies, there have also unfortunately been some disappointments as well. One such disappointment has been the aim to provide a large amount of embryonic stem cells from hybrids.

The Challenge Favourite

To successfully use hybrid stem cells, there are genes that are vital to stem cell development. These genes have to be turned on but thus far, researchers have been unable to accomplish this task. At present, embryonic stem cells are only attainable through discarded embryos left from in vitro fertilisation procedures. They are highly controversial and the search for alternate stem cell sources is a major one, namely to avoid the controversy. While there have also been stories of success in 'reprogramming' adult stem cells into an embryonic-like state, there are still problems with this source.

Embryonic stem cells are ones that have the ability to form any specialised cell in the body while other kinds of stem cells-although still valuable-don't have quite the same potential.

Using Hybrids Favourite

One hope was that by taking the nucleus out of an animal egg and then using the nucleus from a cell in an adult human, we could essentially gain an unlimited supply of embryonic stem cells. The hope had been that the egg's DNA would cause the nucleus to go back to an embryronic state and allow us to extract stem cells that would contain the DNA of the adult who donated the original cells. This would also avoid the risk of immunological rejection in the adult donor, which is another benefit of this therapy if it worked.

Looking at Genes Favourite

In experiments, however, it has unfortunately been shown that genes do not behave as they should in the hybrids. They did not mimic the expression of those genes in normal embryos from humans. Another problem was that the hybrids didn't develop properly to the important stage known as the blastocyst. Here, there is a cluster of approximately a hundred cells, which is vital for obtaining stem cells.

In the hybrids, development seemed to halt early on, generally around the eight to sixteen cell stage. The alerts that are necessary to trigger continued cell development just didn't seem to be there and the nucleus wasn't totally reprogrammed as it should have been. Other scientists, however, are still hopeful that some of these vital genes may eventually be turned on. Scientists in Asia were apparently able to create cybrids from human skin cells and they reported that the blastocyst stage was indeed reached and the critical genes were also turned on. The downside was that the number of cybrids that were able to reach this stage were sparse in number.

Positive Research Favourite

Despite the disappointments of some stem cell research, the future does still look very exciting for finding stem cell treatments that can improve and save lives. Most of the research in this area has been promising and shows that we are making progress into this important area. The setbacks only mean that we must continue to fund ethical stem cell research to search for safe, successful ways to treat disease.

MS and Stem Cells

Along with aiding research into tissue replacement, experts have also developed ways in which stem cells can be used in the treatment of Multiple Sclerosis. This research is not commonly available as a treatment in the UK as trials and experiments continue, though it has been offered in other countries to a handful of patients. In the past, this treatment was offered in Holland, though clinics have been advised not to continue their practices of this treatment as there were many ethical questions that arose from the costs to the source of the stem cells.

Multiple Sclerosis Favourite

Multiple Sclerosis, or MS, is a disorder of the nervous system that can cause loss of balance and can cause loss of sensation in the limbs. Tiredness, visual disturbances and problems with co-ordination are all common symptoms of MS though the degree of severity can be very variable between patients. MS occurs as the protective sheath (the myelin sheath) around the nerve fibres becomes damaged, depleted or scarred. The actual cause of the illness is not fully understood; genetics and external factors are thought to play a role.

Treatment of MS with Stem Cell Favourite

As stem cells developed from either adult or embryonic

cells can grow to be just about any type of cell in the body, they can therefore be developed into nerve cells and those that make up myelin. The process may be carried out by injecting the affected person with stem cells obtained from their own bone marrow, with the aim of targeting the affected areas and replacing or replenishing the depleted and damaged cells. Scientists are finding out extremely valuable information during the research process and are hoping to gain a greater understanding of the disease processes involved with MS along with discovering ethical and suitable treatment options.

Treatment Availability Favourite

As research is still in its infancy in the UK and experts are being extremely careful with the research, this treatment is unlikely to be routinely offered to all sufferers of MS until long term results and effects have been concluded from the trial group. However, there may be the possibility of sufferers volunteering themselves for trials and research with many of the organisations developing this treatment and if this is an option to be considered, the affected person should contact research institutes directly following discussion with your specialist or GP. Research must continue to perfect the treatment and also discover any long term affects such as unwanted cell growth, long term problems and interactions, and whether the treatment is cost effective.

Research in to both stem cell treatments and MS continue every day with new breakthroughs and leads showing promise from the start, but until all trials have been thoroughly tested and concluded, treatments cannot become available for routine use. Treating MS with stem cells, has so far shown quite hopeful results, so may be in a few years, sufferers can look forward to finally finding relief from this debilitating illness. Doctors and the public alike will have to follow developments closely to see what the next instalment of the research includes.

New Harvesting Techniques

New harvesting techniques are crucial to successful stem cell research because they provide greater opportunities to treat diseases in a more unique case-by-case basis. They also provide ways to overcome challenges with current techniques as well as extending stem cell therapies to diseases that may otherwise have been untreatable by current therapies. Perhaps even more important they offer a possible solution to the ethical conundrum that continues to plague stem cell research.

Altered Nuclear Transfer Favourite

Altered nuclear transfer (ANT) may make it feasible for stem cells to be removed from embryos without destroying the embryo itself in the process. This could offer a way out of the current ethical dilemmas surrounding embryonic stem cell research. Normally, embryonic stem cells have been obtained by growing a human embryo into a cluster of cells that contain an inner cell mass. It is the inner cell mass that is abundant in the embryonic stem cells; after this inner cell mass is removed, the embryo is ultimately destroyed.

Traditionally, embryonic stem cells were harvested by destroying the human embryo in a process called somatic cell nuclear transfer (SCNT). A somatic cell is simply a body cell that is neither an egg nor a sperm cell. In this procedure, the nucleus is removed from a somatic cell and it is then implanted into a donor egg that first had its nucleus removed. The egg cell is essentially fooled into thinking it has been fertilized. It has its own DNA and after stimulation, it divides just as a normally fertilized egg would, before forming an embryo. Cells from the inner cell mass are extracted and cultured to provide embryonic stem cells but the technique destroys the embryo. This process has, of course, generated much debate because the embryo could potentially have become a living person. ANT, however, prevents an embryo from actually being created. The nucleus of the somatic cell is altered, or genetically reprogrammed, prior to being transferred into the egg. The alteration consequence is that the somatic cell DNA still produces stem cells but does not generate an embryo.

Blastomere Extraction Favourite

This technique is one potential way around the ethical

concerns that result from the destruction of an embryo. It is performed on a two-day old embryo, following the division of the fertilized egg into eight blastomeres, or cells. Previously, the techniques used for harvesting involving the derivation of embryonic stem cells at a later developmental stage, when the embryo is made up of approximately 150 cells. When these cells were harvested, the embryo was destroyed.

It was found that embryonic stem cells could be extracted from blastomeres, which thus avoids the destruction of the embryo. The cell could be triggered to divide and the resulting stem cells could still be used for research and disease treatment. Further strengthening the research for this technique is the fact that in fertility clinics, the blastomere is already often removed for diagnostic tests to detect genetic abnormalities. The embryo, now with only seven blastomeres can still be implanted into the mother, assuming no defect has been found. These embryos then grow into healthy babies.

Neither ANT nor blastomere extraction are perfect techniques but they do appease many of the ethical concerns surrounding stem cells and can pave the way for new techniques that may one day treat a disease that afflicts you or a loved one.

Peripheral Blood Stem Cell Harvest

A peripheral blood stem cell harvest is a technique used to restore a person's blood cells after they have been damaged by chemotherapy or radiation. The procedure is often used to treat patients with either leukaemia or lymphoma cancer. Because the chemotherapy or radiation treatment damages healthy cells alongside cancer cells a patient requires a viable source of blood-forming cells. Stem cells are able to generate the white blood cells, platelets and red blood cells that are important for functions such as oxygen transport, clotting and immunity.

There are three types of transplants that may be performed for a peripheral harvest:

- *Autologous*: A patient receives his or her own stem cells.

- *Allogeneic*: A patient receives stem cells from someone else-either a relative or an unrelated donor.
- *Syngeneic*: A patient receives stem cells from an identical twin.

The stem cells utilised in this procedure are obtained from the bloodstream in a process called apheresis. In the preceding week before apheresis is performed, a donor receives drugs to increase the number of stem cells in his or her bloodstream. During apheresis, the donor's blood is removed, usually through the arm, and it flows through a machine that removes the stem cells. The blood then flows back to the donor while the extracted stem cells are then frozen until they are transferred to the recipient. The process usually lasts approximately five hours.

Transplantation of Pheripheral Blood Stem Cells Favourite

The patient is treated with either radiation or strong cancer drugs, before receiving stem cells through an intravenous line. After the stem cells are transplanted into the patient, they move from the bloodstream to the bone marrow. It is here that they produce healthy white blood cells, red blood cells and platelets. The process is called engraftment and it generally occurs over the two to four weeks following stem cell transplantation. Physicians will follow a patient's progress by observing blood counts on a regular basis to ensure that the engraftment is successful. A complete recovery is much more long-term, however, and immune functioning can take months to years before it is fully recovered. Sometimes, bone marrow aspiration is performed, where a physician samples a small amount of bone marrow to provide a detailed assessment of the patient's progress.

Advantages of Peripheral Blood Stem Cell Harvestfavourite

Although traditionally, all stem cell transplants that were initially performed were bone marrow transplants, peripheral blood stem cell transplants are now commonly used. It is often the case that more stem cells can be harvested from peripheral blood as opposed to bone marrow. In addition, the procedure for extracting cells from the donor is less taxing and complicated than a bone marrow stem cell harvest. Even more beneficial is the fact that the recipient's blood count tends to replenish much more quickly in comparison with a bone marrow harvest. Therapies such as peripheral blood stem cell harvests are often life-saving. For those who have suffered from cancer or know someone who has battled the disease, stem cell treatments have provided a better chance of remission from the disease. Further research into new treatments as well as improving older ones will ensure that we stay on the path to a cure for cancer.

The Challenges of Stem Cell Therapy

Stem cell therapy encompasses new technologies and therapies that aim to replace damaged cells with healthy new ones. Cells may be dysfunctional due to any number of reasons such as genetics, disease, injury or aging. Currently, stem cells offer the potential to treat cancer, Parkinson's disease, spinal cord injuries and diabetes, among other serious diseases. Unfortunately, there are several challenges faced by researchers that must be overcome before stem cell therapies can become a successful reality for those suffering from disease. Researchers do expect to eventually move beyond these challenges but the unfortunate reality is that those suffering from disease often have little time to wait for treatment.

Identifying Stem Cells in Adult Tissues Favourite

A major difficulty that scientists continue to encounter is the identification of stem cells in adult tissues. These tissues contain many different types of cells and an attempt to locate the often scarce numbers of stem cells in tissues that could contain thousands of different cells is difficult at best.

The research involved is complex and even after cells are isolated, the process to successfully trigger differentiation into the desired cell type is another challenge for researchers. This requires an understanding of stem cell control and regulation that has yet to be fully gained. In addition, researchers must also use the correct laboratory medium, or solution, to coax the growth and this has proven to be difficult.

Stem Cell Integration Favourite

If scientists do manage to identify, isolate and trigger the appropriate differentiate of stem cells, the cells still must be implanted into the patient and accepted among the native body cells. This success is therefore dependent on effective integration into the patient's body systems and other cells.

For example, if cardiac cells are implanted, they must be able to beat in sync with the patient's own heart cells. For a patient who suffers from a neural based disease, any neural cells must integrate into the complicated network of natural neural cells if they are to effectively function and replace damaged cells.

Immunological Rejection Favourite

Immunological rejection is a major barrier to successful stem cell transplants. When a patient's immune system views the transplanted cells as 'foreign' rather than 'self,' it wages an attack on the newly transplanted cells. Recipients of the transplant usually have to take strong immunosuppressive drugs to reduce the chances of rejection but these drugs then leave the patient vulnerable to infection by any other viruses or microbes in the environment.

Cancer Favourite

As scientists learn more and more about stem cells, they hope one day to develop effective cancer treatments based on stem cell research. However, stem cells also have the ability to foster tumour growth and caution must be taken before stem cell therapy becomes commonplace. Cancer is generally a consequence that occurs when cells abnormally divide.

The challenge for scientists is to find a balance between directing cell growth into specialized tissues that can replace damaged ones, and also ensuring that cells don't excessively grow and become cancer cells. Despite the numerous challenges facing researchers, stem cells still hold promise for treating many diseases. Continued research should yield more information to ensure that stem cells treat disease with safety and success.

Therapeutic Cloning

When people think of the word 'cloning' they are often hit with frightening images of duplicate human beings being created in somewhat of a mad scientist style experiment. In fact, many members of the public were outraged when Dolly the sheep resulted from a cloning experiment in Scotland.

Therapeutic cloning, however, is entirely different and does not involve the creation of a perfectly copied human being. It is reproductive cloning that results in a copy of a specific human being. In therapeutic cloning, no sperm fertilisation is involved nor is there implantation into the uterus to create a child.

Therapeutic cloning is another phrase for a procedure known as somatic cell nuclear transfer (SCNT). In this procedure, a researcher extracts the nucleus from an egg. The nucleus holds the genetic material for a human or laboratory animal. Scientists then take a somatic cell, which is any body cell other than an egg or sperm, and also extract the nucleus from this cell. In practical human applications, the somatic cell would be taken from a patient who requires a stem cell transplant to treat a health condition or disease.

The nucleus that is extracted from the somatic cell in the patient is then inserted into the egg, which had its nucleus previously removed. In a very basic sense, it's a procedure of substitution. The egg now contains the patient's genetic material, or instructions. It is stimulated to divide and shortly thereafter forms a cluster of cells known as a blastocyst. This blastocyst has both an outer and inner layer of cells and it is the inner layer, called the inner cell mass that is rich in stem cells.

The cells in the inner cell mass are isolated and then utilised to create embryonic stem cell lines, which are infused into the patient where they are ideally integrated into the tissues, imparting structure and function as needed.

Benefits of Therapeutic Cloning Favourite

A major benefit of therapeutic cloning is that the cells removed are pluripotent. Pluripotent cells can give rise to all cells in the body with the exception of the embryo. This means that pluripotent cells can potentially treat diseases in any body organ or tissue by replacing damaged and dysfunctional cells.

Another distinct advantage to this type of therapy is that the risk of immunological rejection is alleviated because the patient's own genetic material is used. If a cell line were created with cells from another individual, the patient's body would be more likely to recognise the foreign proteins and then wage an attack on the transplanted cells. The ultimate consequence would be a rejected stem cell transplant. This is one of the major challenges of organ transplants, alongside the fact that there is a huge shortage of available organs for those who require the procedure.

This means that therapeutic cloning has the potential to dramatically reduce the wait times for organ transplants as well as the immunological concerns associated with organ transplant therapy. Therapeutic cloning is also important to enhancing our understanding of stem cells and how they and other cells develop. This understanding can hopefully lead to new treatments or cures for some of the common diseases affecting people today. In addition, the procedure would allow for scientists to create stem cell therapies that are patient specific and perfectly matched for the patient's medical condition.

Problems with Therapeutic Cloning Favourite

One problem with therapeutic cloning is that many attempts are often required to create a viable egg. The stability of the egg with the infused somatic nucleus is poor and it can require hundreds of attempts before success is attained. Therapeutic cloning does result in the destruction of an embryo after stem cells are extracted and this destruction has stirred controversy over the morality of the procedure. Some argue that the pros outweigh the cons with regards to treating disease whilst others have likened the destruction to an abortion. Still others state that this doesn't change the fact the embryo could potentially be a human being and so destruction of the embryo is no different than destruction of a human life. Because reproductive cloning does utilise SCNT as the primary step, there is also still fear that given our knowledge base to perform reproductive cloning, a scientist may attempt to move beyond therapeutic cloning to creation of a human being.

To this date, no human being has been successfully cloned but the possibility of this occurring is a frightening one not only for the general public and policy makers, but also for most of the ethical scientific field. The majority of scientists are adamantly opposed to reproductive cloning and instead, support therapeutic cloning for treating disease. With policies and careful monitoring in place to ensure that therapeutic cloning is used responsibly, we can all benefit from the potential of this procedure to eventually treat, or perhaps one day cure, many diseases.

12

Culture of Microbial Cells

Understanding modern molecular cell biology requires a familiarity with the most commonly used biological materials and knowledge of the latest experimental techniques. In this chapter, we describe the primary techniques used to culture and manipulate certain cells, viruses, and experimental organisms. Biologists choose to study these cells and organisms because they provide experimentally favourable examples of important molecular events or processes such as the control of gene activity and cell replication, formation of organelles, secretion of proteins, and differentiation of cells.

In the first part of the chapter, we discuss the culture of microbial cells, animal cells, and hybrid cells in vitro and give some examples of their experimental uses. Many studies involve cells or organisms with mutations affecting a specific biochemical or developmental pathway. In such studies, an important criterion for using a particular cell type or organism is the ease with which mutants can be isolated and characterized and the mutated genes identified and cloned. Because viruses contain a small number of genes (4 to ≈100 in different types of viruses), viral mutants commonly are used to study the action of a limited set of genes designed to carry out restricted, specific molecular tasks. To provide a basis for understanding such studies, we consider the basic properties of viruses and their uses in various types of research in the latter part of this chapter.

GROWTH OF MICROORGANISMS IN CULTURE

Cultured cells have several advantages over intact

organisms for research on fundamental aspects of cell biology. First, most animal and plant tissues comprise a variety of different types of cells, whereas cells of a specific type can be grown in culture; thus cultured cells are more homogeneous in their properties than in vivo cells in tissues. Second, experimental conditions can be controlled much more rigorously in culture than in an organism.

By manipulating the growth conditions, for example, one can measure the effects of specific chemicals and growth factors on a particular cell type in culture. A third reason for preferring cultured cells is that in many cases a single cell can be readily grown into a colony, a process called *cell cloning,* or simply *cloning*. The resulting strain of cells, which is genetically homogeneous, is called a clone. This simple technique, which is commonly used with many bacteria, yeasts, and mammalian cell types, makes it easy to isolate genetically distinct clones of cells.

GROWTH OF MICROORGANISMS IN MINIMAL MEDIUM

Among the advantages of using microorganisms such as the bacterium *Escherichia coli* and the yeast *Saccharomyces cerevisiae* are their rapid growth rate and simple nutritional requirements, which can be met with a minimal medium. A minimal medium for such microorganisms can contain glucose as the sole source of carbon; metabolism of glucose to smaller molecules (*e.g.*, CO_2, ethanol, or acetic acid) can generate the ATP necessary for energy-requiring activities of the cells. The sole nitrogen source in a minimal medium can be ammonium (NH_4^+), from which the cells can synthesize all the necessary amino acids and other nitrogen-containing metabolites. Salts and trace elements are the only other components of a minimal medium.

Many prokaryotes (*i.e.*, bacteria) and single-celled eukaryotes such as yeast, both of which grow in nature as single cells, are easily grown in culture dishes — usually on top of *agar,* a semisolid base of plant polysaccharides. The agar is first dissolved in a heated nutrient medium, and the solution is poured into petri dishes; as it cools, the solution solidifies. A dilute suspension of cells then is dispersed on top of the agar; in time each cell grows into a discrete colony. Since the cells in a colony all derive from a single cell, they form a clone and have identical genomes (DNA). All the cells in a clone generally express the same set of genes and contain the same enzymes and other constituents in similar proportions. The division time for *E. coli* and similar microorganisms ranges from 20 minutes to 1 hour. Thus a single *E. coli* cell, which divides approximately every 30 minutes, can grow into a colony containing $10^7 - 10^8$ cells in 12 hours ($2^{24} = 1.7 \times 10^7$).

Because yeasts, unlike most eukaryotic organisms, grow as single cells and can be grown in a simple defined medium similarly to bacteria, they are a popular choice for studies of eukaryotic cell function. The entire life cycle of yeasts can be studied in culture, and colonies can be grown from a single vegetative cell (a growing cell) or from a single spore (a dormant cell). Studies with yeasts have provided valuable information on such subjects as the cellular mechanisms for controlling DNA synthesis, genetic recombination, and the gene products that regulate the cycle of events leading to replication and segregation of chromosomes and ultimately cell division.

ISOLATION OF BACTERIA AND YEAST BY REPLICA PLATING

If a random mutation in a microorganism cultured, or plated, on a minimal agar medium renders cells unable to synthesize a necessary metabolite, and this metabolite is not supplied in the medium (*e.g.*, the amino acid arginine), then the cells will die. But if the cells are grown on a rich medium that supplies the metabolite whose synthesis is prevented by a mutation, then the mutant cell can survive and grow into a colony in which all the cells have the same mutation.

A mutant colony, say, an Arg″ colony, can be detected by *replica plating*, a technique developed by Joshua and Esther Lederberg in the 1940s. The initial plate of multiple colonies growing on rich medium is called the *master plate*. A circular, velvet-covered stamp equal in size to the master plate is pressed onto it, so that the stamp picks up cells from each colony.

The stamp is then used to deposit cells onto a second plate containing minimal agar medium without arginine (Arg$^{\prime\prime}$ medium) and also onto a third plate containing minimal agar medium with arginine (Arg^{+} medium). Cells that can synthesize arginine grow into colonies on both the Arg$^{\prime\prime}$ and Arg^{+} plates.

However, a clone of cells on the master plate with a mutation in a gene required for arginine synthesis cannot grow on the Arg$^{\prime\prime}$ medium, resulting in a space on the Arg$^{\prime\prime}$ plate corresponding to the position of this clone on the master plate and Arg^{+} plate. This clone of cells, thus identified as being defective in the synthesis of arginine, then can be isolated from the master plate.

Repeated replica plating of a master plate with paired minimal-medium plates containing or lacking different specific nutrients (*e.g.*, amino acids, nucleic acid bases, vitamins) can identify clones of cells with mutations in the genes required for synthesis of different biological molecules. This approach has been used to isolate thousands of different bacterial and yeast nutritional auxotrophs; these are strains that are unable to grow in the absence of a specific nutrient.

Under standard culture conditions, random mutations occur in *E. coli* genes at a frequency of 1 in 10^6 – 10^8 base pairs per generation. Thus, by the time a single cell has divided to produce a few hundred or thousand cells, some individual genes will differ among cells in the clone. To maintain the genetic homogeneity of a clone, it must be recloned frequently under selective conditions that permit only cells with a particular set of desired characteristics to survive and grow. For example, to maintain an Arg$^{\prime\prime}$ clone, isolated as described above, the cells are replated on minimal medium plus arginine; only cells that can synthesize all the metabolites required for growth except arginine will grow on this selective medium.

GROWTH OF ANIMAL CELLS IN CULTURE

Animal cells are more difficult to culture than microorganisms because they require many more nutrients and typically grow only when attached to specially coated surfaces. Despite these difficulties, various types of animal cells, including both undifferentiated and differentiated ones, can be cultured successfully.

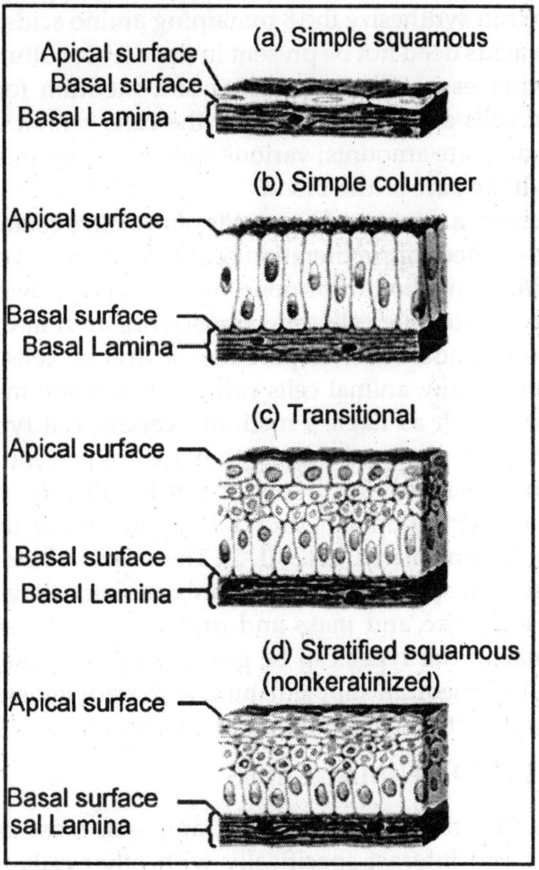

Fig. Principal types of epithelium.

Nine amino acids, referred to as the *essential amino acids,* cannot be synthesized by adult vertebrate animals and thus must be obtained from their diet. Animal cells grown in culture also must be supplied with these nine amino acids, namely, histidine, isoleucine, leucine, lysine, methionine, phenylalanine, threonine, tryptophan, and valine. In addition, most cultured cells require cysteine, glutamine, and tyrosine. In the intact animal, these three amino acids are synthesized by specialized cells; for example, liver cells make tyrosine from phenylalanine, and both liver and kidney cells can make glutamine. Animal cells both within the organism and in culture can synthesize the 8 remaining amino acids; thus these amino acids need not be present in the diet or culture medium. The other essential components of a medium for culturing animal cells are vitamins, which the cells cannot make at all or in adequate amounts; various salts; glucose; and *serum,* the noncellular part of the blood.

Serum, a mixture of hundreds of proteins, contains various factors needed for proliferation of cells in culture. For example, it contains insulin, a hormone required for growth of many cultured vertebrate cells, and transferrin, an iron-transporting protein essential for incorporation of iron by cells in culture. Although many animal cells can grow in a serum-containing medium, such as Eagle's medium, certain cell types require specific protein growth factors that are not present in serum.

For instance, precursors of red blood cells require the hormone erythropoietin, and T lymphocytes of the immune system require interleukin 2 (IL-2). These factors bind to receptor proteins that span the plasma membrane, signaling the cells to increase in size and mass and undergo cell division. A few mammalian cell types can be grown in a completely defined, serum-free medium supplemented with trace minerals, specific protein growth factors, and other components.

CULTURED ANIMAL CELLS

Within the tissues of intact animals, most cells tightly contact and interact specifically with other cells via various cellular junctions. The cells also contact the extracellular matrix, a complex network of secreted proteins and carbohydrates that fills the spaces between cells. The matrix, whose constituents are secreted by cells themselves, helps bind the cells in tissues together; it also provides a lattice through which cells can move, particularly during the early stages of animal differentiation.

The extracellular matrices in various animal tissues consist of several common components: fibrous collagen proteins; hyaluronan (or hyaluronic acid), a large mucopolysaccharide; and covalently linked polysaccharides and proteins in the form of proteoglycans (mostly carbohydrate) and glycoproteins (mostly protein). However, the exact composition of the matrix in different tissues varies, reflecting the specialized function of a tissue. In connective tissue, for example, the major protein of the extracellular matrix is a type of collagen that forms insoluble fibres with a very high tensile strength. Fibroblasts, the principal cell type in connective tissue, secrete this type of collagen as well as the other matrix components. Receptor proteins in the plasma membrane of cells bind various matrix elements, imparting strength and rigidity to tissues.

The tendency of animal cells in vivo to interact with one another and with the surrounding extracellular matrix is mimicked in their growth in culture. Unlike bacterial and yeast cells, which can be grown in suspension, most cultured animal cells require a surface to grow on. Many types of cells can adhere to and grow on glass, or on specially treated plastics with negatively charged groups on the surface (*e.g.*, SO_3^{2-}). The cultured cells secrete collagens and other matrix components; these bind to the culture surface and function as a bridge between it and the cells. Cells cultured from single cells on a glass or a plastic dish form visible colonies in 10 – 14 days. Some tumor cells can be grown in suspension, a considerable experimental advantage because equivalent samples are easier to obtain from suspension cultures than from colonies grown in a dish.

PRIMARY CELL CULTURES

Normal animal tissues (*e.g.*, skin, kidney, liver) or whole embryos commonly are used to establish *primary cell cultures.* To prepare tissue cells for culture (or to remove adherent cells from a culture dish for biochemical studies), trypsin or another protease is used to destroy the proteins in the junctions that normally interconnect cells. For many years, most cell types were difficult, if not impossible, to culture. But the identification and preparation of various protein growth factors that stimulate the replication of specific cell types, as well as other recent modifications in culture methods, now permit experimenters to grow various types of specialized cells.

Many studies with vertebrate cells, however, still are performed with those few cell types that grow most readily in culture. These are not cells of a defined type; rather, they represent whatever grows when a tissue or an embryo is placed in culture. The cell type that usually predominates in such cultures is called a fibroblast because it secretes the types of proteins associated with fibroblasts in fibrous connective tissue of animals. Cultured fibroblasts have the morphology of tissue fibroblasts, but they retain the ability to differentiate into other cell types; thus they are not as differentiated as tissue fibroblasts.

Some studies are conducted with primary cultures of epithelial cells. In general, external and internal surfaces of tissues and organs are covered by a layer of epithelial cells called an epithelium. These highly differentiated cells are said to be *polarized* because the plasma membrane is organized into at least two discrete regions. For example, the epithelial cells that line the intestine form a simple columnar epithelium. That portion of the plasma membrane facing the lumen of the intestine, the *apical* surface, is specialized for absorption; the rest of the plasma membrane, the *basolateral* surface, mediates transport of nutrients from the cell to the blood and forms junctions with adjacent cells and the underlying extracellular matrix called the basal lamina.

Certain cells cultured from blood, spleen, or bone marrow adhere poorly, if at all, to a culture dish but nonetheless grow well. In the body, such nonadherent cells are held in suspension (in the blood), or they are loosely adherent (in the bone marrow and spleen). Because these cells often come from immature stages in the development of differentiated blood cells, they are very useful for studying normal blood cell differentiation and the abnormal development of leukemias.

When cells are removed from an embryo or an adult animal, most of the adherent ones grow continuously in culture for only a limited time before they spontaneously cease growing. Such a culture eventually dies out after many cell doublings, even if it is provided with fresh supplies of all the known nutrients that cells need to grow, including serum. For instance, when human fetal cells are explanted into cell culture, the majority of cells die within a relatively short time; "fibroblasts," although also destined to die, proliferate for a while and soon become the predominant cell type. They divide about 50 times before they cease growth. Starting with 10^6 cells, 50 doublings can produce $10^6 \times 2^{50}$, or more than 10^{20} cells, which is equivalent to the weight of about 10^5 people. Thus, even though its lifetime is limited, a single culture, if carefully maintained, can be studied through many generations. Such a lineage of cells originating from one initial primary culture is called a cell strain.

TRANSFORMED CELLS CAN GROW INDEFINITELY IN CULTURE

To be able to clone individual cells, modify cell behaviour, or select mutants, biologists often want to maintain cell cultures for many more than 100 doublings. This is possible with cells derived from some tumors and with rare cells that arise spontaneously because they have undergone genetic changes that endow them with the ability to grow indefinitely. The genetic changes that allow these cells to grow indefinitely are collectively called *oncogenic* transformation, and the cells are said to be *oncogenically transformed,* or simply *transformed.* A culture of cells with an indefinite life span is considered immortal; such a culture is called a cell line to distinguish it from an impermanent cell *strain.*

The ability of cultured cells to grow indefinitely or their tendency to be transformed varies depending on the animal species from which the cells originate. Normal chicken cells rarely are transformed and die out after only a few doublings;

even tumor cells from chickens almost never exhibit immortality. Among human cells, only tumor cells grow indefinitely. The HeLa cell, the first human cell type to be grown in culture, was originally obtained in 1952 from a malignant tumor (carcinoma) of the uterine cervix. This cell line has been invaluable for research on human cells.

In contrast to human and chicken cells, cultures of embryonic adherent cells from rodents routinely give rise to cell lines. When adherent rodent cells are first explanted, they grow well, but after a number of serial replatings they lose growth potential and the culture goes into crisis. During this period most of the cells die, but often a rapidly dividing variant cell arises spontaneously and takes over the culture. A cell line derived from such a variant will grow forever if it is provided with the necessary nutrients. Cells in spontaneously established rodent cell lines and in cell lines derived from tumors often have abnormal chromosomes. In addition, their chromosome number usually is greater than that of the normal cell from which they arose, and it continually expands and contracts in culture. Such cells are said to be *aneuploid* (*i.e.*, have an inappropriate number of chromosomes) and are obviously mutants.

Although most cell lines are undifferentiated, some can carry out many of the functions characteristic of the normal differentiated cells from which they are derived. One example is certain hepatoma cell lines (*e.g.*, HepG2) that synthesize most of the serum proteins made by normal hepatocytes (the major cell type in the liver) from which they are derived. These highly differentiated hepatoma cells are often studied as models of normal hepatocytes. Cultured *myoblasts* (muscle precursor cells) are another example of transformed cells that continue to perform many functions of a specialized, differentiated cell. When grown in culture, transformed myoblasts can be induced to fuse to form myotubes. These resemble differentiated multinucleated muscle cells and synthesize many of, if not all, the specialized proteins associated with contraction. Certain lines of epithelial cells also have been cultured successfully. One such line, Madin-Darby canine kidney (MDCK) cells, forms a continuous sheet of polarized epithelial cells one cell thick that exhibits many of the properties of the normal canine kidney epithelium from which it was derived. This type of preparation has proved valuable as a model for studying the functions of epithelial cells.

FUSION OF CULTURED ANIMAL CELLS

Cultured animal cells infrequently undergo cell fusion spontaneously. The fusion rate, however, increases greatly in the presence of certain viruses that have a lipoprotein envelope similar to the plasma membrane of animal cells. A mutant viral glycoprotein in the envelope promotes cell fusion the mechanism of this effect is discussed. Cell fusion also is promoted by polyethylene glycol, which causes the plasma membranes of adjacent cells to adhere to each other and to fuse. As most fused animal cells undergo cell division, the nuclei eventually fuse, producing viable cells with a single nucleus that contains chromosomes from both "parents." The fusion of two cells that are genetically different yields a hybrid cell called a heterokaryon.

Because some somatic cells from animals can be cultured from single cells in a well-defined medium, it is possible to select for genetically distinct cultured animal cells, just as is done with bacterial and yeast cells. Moreover, during mitosis the chromosomes in an animal cell are large and highly visible after staining, making it easy to distinguish individual chromosomes. Genetic studies of cultured animal cells are called *somatic-cell genetics* to distinguish them from *classical genetics,* which deals with whole organisms derived from germ cells (sperm and eggs).

Cultured cells from different mammals can be fused to produce interspecific hybrids, which have been widely used in somatic-cell genetics. For instance, hybrids can be prepared from human cells and mutant mouse cells that lack an enzyme required for synthesis of a particular essential metabolite. As the human-mouse hybrid cells grow and divide, they gradually lose human chromosomes in random order, but retain the mouse chromosomes. In a medium that can support growth of both the human cells and mutant mouse cells, the hybrids eventually lose all human chromosomes. However, in a medium lacking the essential metabolite that the mouse cells cannot produce, the one human chromosome that contains the gene encoding the needed enzyme will be retained, because any hybrid cells that lose it following mitosis will die. All other human chromosomes eventually are lost.

By using different mutant mouse cells and media in which they cannot grow, researchers have prepared various panels of hybrid cell lines. Each cell line in a panel contains either a single human chromosome or a small number of human chromosomes, and a full set of mouse chromosomes. Because each chromosome can be identified visually under a light microscope, such hybrid cells provide a means for assigning, or "mapping," individual genes to specific chromosomes. For example, suppose a hybrid cell line is shown microscopically to contain a particular human chromosome. That hybrid cell line can then be tested biochemically for the presence of various human enzymes, exposed to specific antibodies to detect human surface antigens, or subjected to DNA hybridization and cloning techniques to locate particular human DNA sequences. The genes encoding a human protein or containing a human DNA sequence detected in such tests must be located on the particular human chromosome carried by the cell line being tested. Panels of hybrids between normal mouse and mutant hamster cells also have been established; in these hybrid cells, the majority of mouse chromosomes are lost, allowing mouse genes to be mapped to specific mouse chromosomes.

HYBRID CELLS

One metabolic pathway has been particularly useful in cell-fusion experiments. Most animal cells can synthesize the purine and pyrimidine nucleotides de novo from simpler carbon and nitrogen compounds, rather than from already formed purines and pyrimidines. The folic acid antagonists amethopterin and aminopterin interfere with the donation of methyl and formyl groups by tetrahydrofolic acid in the early stages of de novo synthesis of glycine, purine nucleoside monophosphates, and thymidine monophosphate. These drugs are called *antifolates,* since they block reactions involving tetrahydrofolate, an active form of folic acid. Many cells, however, contain enzymes that can synthesize the necessary nucleotides from purine bases and thymidine if they are provided in the medium; these *salvage pathways* bypass the metabolic blocks imposed by antifolates.

A number of mutant cell lines lacking the enzyme needed to catalyze one of the steps in a salvage pathway have been isolated. For example, cell lines lacking thymidine kinase (TK) can be selected because such cells are resistant to the otherwise toxic thymidine analog 5-bromodeoxyuridine. Cells containing TK convert 5-bromodeoxyuridine into 5-bromodeoxyuridine monophosphate. This nucleoside mono- phosphate is then converted into a nucleoside triphosphate by other enzymes and is incorporated by DNA polymerase into DNA, where it exerts its toxic effects. This pathway is blocked in cells with a *TK* mutation that prevents production of functional TK enzyme. Hence, TK″ mutants are resis- tant to the toxic effects of 5-bromodeoxyuridine. Similarly, cells lacking the HGPRT enzyme have been selected because they are resistant to the otherwise toxic guanine analog 6-thioguanine. As we will see next, HGPRT″ cells and TK″ cells are useful partners in cell fusions with one another or with cells that have salvage-pathway enzymes but that are differentiated and cannot grow in culture by themselves.

The medium most often used to select hybrid cells is called *HAT medium,* because it contains hypoxanthine (a purine), aminopterin, and thymidine. Normal cells can grow in HAT medium because even though aminopterin blocks de novo synthesis of purines and TMP, the thymidine in the media is transported into the cell and converted to TMP by TK and the hypoxanthine is transported and converted into usable purines by HGPRT. On the other hand, neither TK″ nor HGPRT″ cells can grow in HAT medium because each lacks an enzyme of the salvage pathway. However, hybrids formed by fusion of these two mutants will carry a normal *TK* gene from the

HGPRT″ parent and a normal *HGPRT* gene from the TK″parent. The hybrids thus will produce both functional salvage-pathway enzymes and grow on HAT medium. Likewise, hybrids formed by fusion of mutant cells and normal cells can grow in HAT medium.

Hybridomas Are Used to Produce Monoclonal Antibodies: Each normal B lymphocyte in an animal is capable of producing a single type of antibody directed against a specific determinant, or epitope, on an antigen molecule. If an animal is injected with an antigen, B lymphocytes that make antibody recognizing the antigen are stimulated to grow and proliferate. Each antigen-activated B lymphocyte forms a clone of cells in the spleen or lymph nodes, with each cell of the clone producing identical antibody, termed monoclonal antibody. Because most natural antigens contain multiple epitopes, exposure of an animal to an antigen usually stimulates formation of several different B-lymphocyte clones, each producing a different antibody; a mixture of antibodies that recognize different epitopes on the same antigen is said to be *polyclonal*.

For many types of studies involving antibodies, monoclonal antibody is preferable to polyclonal antibody. However, biochemical purification of monoclonal antibody from serum is not feasible, in part because the concentration of any given antibody is quite low. For this reason, researchers looked to culture techniques in order to obtain usable quantities of monoclonal antibody. Because primary cultures of normal B lymphocytes do not grow indefinitely, such cultures have limited usefulness for production of monoclonal antibody. This limitation can be avoided by fusing normal B lymphocytes with oncogenically transformed lymphocytes called *myeloma cells*, which are immortal.

Fusion of a myeloma cell with a normal antibody-producing cell from a rat or mouse spleen yields a hybrid that proliferates into a clone called a hybridoma. Like myeloma cells, hybridoma cells are immortal. Each hybridoma produces the monoclonal antibody encoded by its B-lymphocyte partner. Many different myeloma cell lines from mice and rats have been established; from these, HGPRT″ lines have been selected based on their resistance to 6-thioguanine as described above. If such mutant myeloma cells are fused with normal B lymphocytes, any fused cells that result can grow in HAT medium, but the parental cells cannot. Each selected hybridoma then is tested for production of the desired antibody; any clone producing that antibody then is grown in large cultures, from which a substantial quantity of pure monoclonal antibody can be obtained.

Such pure antibodies are very valuable research reagents. For example, a monoclonal antibody that interacts with protein X can be used to label, and thus locate, protein X in specific cells of an organ or in specific cell fractions. Once identified, even very scarce proteins can be isolated by affinity chromatography in columns to which the monoclonal antibody is bound. Monoclonal antibodies also have become important diagnostic and therapeutic tools in medicine. Monoclonal antibodies that bind to and inactivate toxic proteins (toxins) secreted by bacterial pathogens are used to treat diseases caused by these pathogens. Other monoclonal antibodies are specific for cell-surface proteins expressed by certain types of tumor cells; chemical complexes of such monoclonal antibodies with toxic drugs are being developed for cancer chemotherapy.

STRUCTURE AND FUNCTION OF VIRUSES

A virus is a small parasite that cannot reproduce by itself. Once it infects a susceptible cell, however, a virus can direct the cell machinery to produce more viruses. Most viruses have either RNA or DNA as their genetic material. The nucleic acid may be single- or double-stranded. The entire infectious virus particle, called a virion, consists of the nucleic acid and an outer shell of protein. The simplest viruses contain only enough RNA or DNA to encode four proteins. The most complex can encode 100 – 200 proteins.

The study of plant viruses inspired some of the first experiments in molecular biology. In 1935, Wendell Stanley purified and partly crystallized tobacco mosaic virus (TMV); other plant viruses were crystallized soon thereafter. Pure proteins had been crystallized only a short time before

Stanley's work, and it was considered very surprising at the time that a replicating organism could be crystallized.

A wealth of subsequent research with bacterial viruses and animal viruses has provided detailed understanding of viral structure, and virus-infected cells have proved extremely useful as model systems for the study of basic aspects of cell biology. In many cases, DNA viruses utilize cellular enzymes for synthesis of their DNA genomes and mRNAs; all viruses utilize normal cellular ribosomes, tRNAs, and translation factors for synthesis of their proteins. Most viruses commandeer the cellular machinery for macromolecular synthesis during the late phase of infection, directing it to synthesize large amounts of a small number of viral mRNAs and proteins instead of the thousands of normal cellular macromolecules.

For instance, animal cells infected by influenza or vesicular stomatitis virus synthesize only one or two types of glycoproteins, which are encoded by viral genes, whereas uninfected cells produce hundreds of glycoproteins. Such virus-infected cells have been used extensively in studies on synthesis of cell-surface glycoproteins. Similarly, much information about the mechanism of DNA replication has come from studies with bacterial cells and animal cells infected with simple DNA viruses, since these viruses depend almost entirely on cellular proteins to replicate their DNA.

Viruses also often express proteins that modify host-cell processes so as to maximize viral replication. For example, the roles of certain cellular factors in initiation of protein synthesis were revealed because viral proteins interrupt their action. Finally, when certain genes carried by cancer-causing viruses integrate into chromosomes of a normal animal cell, the normal cell can be converted to a cancer cell.

Since many viruses can infect a large number of different cell types, genetically modified viruses often are used to carry foreign DNA into a cell. This approach provides the basis for a growing list of experimental gene therapy treatments. Because of the extensive use of viruses in cell biology research and their potential as therapeutic agents, we describe the basic aspects of viral structure and function in this section.

VIRAL CAPSIDS

The nucleic acid of a virion is enclosed within a protein coat, or capsid, composed of multiple copies of one protein or a few different proteins, each of which is encoded by a single viral gene. Because of this structure, a virus is able to encode all the information for making a relatively large capsid in a small number of genes. This efficient use of genetic information is important, since only a limited amount of RNA or DNA, and therefore a limited number of genes, can fit into a virion capsid. A capsid plus the enclosed nucleic acid is called a nucleocapsid.

Nature has found two basic ways of arranging the multiple capsid protein subunits and the viral genome into a nucleocapsid. The simpler structure is a protein helix with the RNA or DNA protected within. Tobacco mosaic virus (TMV) is a classic example of the helical nucleocapsid. In TMV the protein subunits form broken disklike structures, like lock washers, which form the helical shell of a long rodlike virus when stacked together.

The other major structural class of viruses, called *icosahedral* or *quasi-spherical viruses,* is based on the icosahedron, a solid object built of 20 identical faces, each of which is an equilateral triangle. In the simplest type of icosahedral virion each of the 20 triangular faces is constructed of three identical capsid protein subunits, making a total of 60 subunits per capsid. At each of the 12 vertices, five subunits make contact symmetrically. Thus all protein subunits are in *equivalent* contact with one another. Tobacco satellite necrosis virus has such a simple icosahedral structure. However, most quasi-spherical viruses are larger, requiring the assembly of more than three subunits per face of the icosahedron. These proteins form shells whose subunits are in *quasi-equivalent* contact. Here, the proteins at the icosahedral vertices remain arranged in a fivefold symmetry, but additional subunits cover the surfaces between in a pattern of sixfold symmetry.

The atomic structures of a number of icosahedral viruses have been determined by x-ray crystallography. The first three such viruses to be analysed—tomato bushy stunt virus, poliovirus, and rhinovirus (the common cold virus) — exhibit a remarkably similar design, in terms of the rules of icosahedral symmetry as well as in the details of their surface proteins. In each virus, at atomic resolution, clefts ("canyons") are observed encircling each of the vertices of the icosahedral structure. Interaction of these clefts with cell-surface receptors attaches the virus to a host cell, the first step in viral infection. Neutralizing antibodies specific for a particular virus also interact with these clefts, thereby inhibiting attachment of the virus to the host cell.

In some viruses, the symmetrically arranged nucleocapsid is covered by an external membrane, or *envelope,* which consists mainly of a phospholipid bilayer but also contains one or two types of virus-encoded glycoproteins. The phospholipids in the viral envelope are similar to those in the plasma membrane of an infected host cell. The viral envelope is, in fact, derived by budding from that membrane, but contains mainly viral glycoproteins.

The components of simple viruses such as TMV, which consists of a single RNA molecule and one protein species, undergo self-assembly if they are mixed in solution. More complex viruses containing a dozen or more protein species do not spontaneously assemble in vitro. The multiple components of such viruses assemble within infected cells in stages, first into subviral particles and then into completed virions. The genomes of these complex viruses encode proteins that assist in the assembly of the virion, but the assembly proteins are not themselves components of the completed virion.

MOST VIRAL HOST RANGES ARE NARROW

The fact that the host range — the group of cell types that a virus can infect — is generally restricted serves as a basis for classifying viruses. A virus that infects only bacteria is called a bacteriophage, or simply a phage. Viruses that infect animal or plant cells are referred to generally as *animal viruses* or *plant viruses.* A few viruses can grow in both plants and the insects that feed on them. The highly mobile insects serve as vectors for transferring such viruses between susceptible plant hosts. An example is potato yellow dwarf virus, which can grow in leafhoppers (insects that feed on potato plant leaves) as well as in potato plants. Wide host ranges are characteristic of some strictly animal viruses, such as vesicular stomatitis virus, which grows in insects and in many different types of mammalian cells. Most animal viruses, however, do not cross phyla, and some (*e.g.*, poliovirus) infect only closely related species such as primates. The host-cell range of some animal viruses is further restricted to a limited number of cell types because only these cells have appropriate surface receptors to which the virions can attach.

VIRUSES CAN BE CLONED

The number of infectious viral particles in a sample can be quantified by a plaque assay. This assay is performed by culturing a dilute sample of viral particles on a plate covered with host cells and then counting the number of local lesions, called *plaques,* that develop. A plaque develops on the plate wherever a single virion initially infects a single cell. The virus replicates in this initial host cell and then lyses the cell, releasing many progeny virions that infect the neighbouring cells on the plate. After a few such cycles of infection, enough cells are lysed to produce a visible plaque in the layer of remaining uninfected cells.

Since all the progeny virions in a plaque are derived from a single parental virus, they constitute a virus clone. This type of plaque assay is in standard use for bacterial and animal viruses. Plant viruses can be assayed similarly by counting local lesions on plant leaves inoculated with viruses. Analysis of viral mutants, which are commonly isolated by plaque assays, has contributed extensively to current understanding of molecular cellular processes. The plaque assay also is critical in isolating λ bacteriophage clones carrying segments of cellular DNA.

LYSOGENIC

The surface of viruses includes many copies of one type

of protein that binds, or adsorbs, specifically to multiple copies of a receptor protein on a host cell. This interaction determines the host range of a virus and begins the infection process. Then, in one of various ways, the viral DNA or RNA crosses the plasma membrane into the cytoplasm. The entering genetic material may still be accompanied by inner viral proteins, although in the case of many bacteriophages, all capsid proteins remain outside an infected cell. The genome of most DNA-containing viruses that infect eukaryotic cells is transported (with some associated proteins) into the cell nucleus, where the cellular DNA is, of course, also found. Once inside the cell, the viral DNA interacts with the host's machinery for transcribing DNA into mRNA. The viral mRNA that is produced then is translated into viral proteins by host-cell ribosomes, tRNA, and translation factors.

Most viral protein products fall into one of three categories: special enzymes needed for viral replication; inhibitory factors that stop host-cell DNA, RNA, and protein synthesis; and structural proteins used in the construction of new virions. These last proteins generally are made in much larger amounts than the other two types. After the synthesis of hundreds to thousands of new virions has been completed, most infected bacterial cells and some infected plant and animal cells rupture, or lyse, releasing all the virions at once. In many plant and animal viral infections, however, no discrete lytic event occurs; rather, the dead host cell releases the virions as it gradually disintegrates.

These events — adsorption, penetration, replication, and release — describe the lytic cycle of viral replication. The outcome is the production of a new round of viral particles and death of the cell. Adsorption and release of enveloped animal viruses are somewhat more complicated processes. In this case, the virions "bud" from the host cell, thereby acquiring their outer phospholipid envelope, which contains mostly viral glycoproteins.

We illustrate the lytic cycle of enveloped viruses with the rabies virus, whose nucleocapsid consists of a single-stranded RNA genome surrounded by multiple copies of nucleocapsid protein. Within the nucleocapsid of rabies virions are viral enzymes for synthesizing viral mRNA and replicating the viral genome. The envelope around the nucleocapsid is a phospholipid bilayer containing multiple copies of a viral transmembrane glycoprotein. This receptor-binding, or "attachment," protein has a large external folded domain on the outside of the viral envelope, an α-helical transmembrane domain that spans the viral envelope, and a short internal domain.

The internal domain interacts with the viral matrix protein, which functions as a bridge between the transmembrane glycoprotein and nucleocapsid protein. Assembly of progeny nucleocapsids, and release of progeny virions by budding from the host-cell plasma membrane. Budding virions are clearly visible in electron micrographs.

In some cases, after a bacteriophage DNA molecule enters a bacterial cell, it becomes integrated into the host-cell chromosome, where it remains quiescent and is replicated as part of the cell's DNA from one generation to the next. This association is called lysogeny, and the integrated phage DNA is referred to as a *prophage*. Under certain conditions, the prophage DNA is activated, leading to its excision from the host-cell chromosome and entrance into the lytic cycle. Bacterial viruses of this type are called *temperate phages*. The genomes of a number of animal viruses also can integrate into the host-cell genome. Probably the most important are the retroviruses, described briefly later in this chapter.

A few phages and animal viruses can infect a cell and cause new virion production without killing the cell or becoming integrated. Four Types of Bacterial Viruses Are Widely Used in Biochemical and Genetic Research

Bacterial viruses have played a crucial role in the development of molecular cell biology. Thousands of different bacteriophages have been isolated; many of these are particularly well suited for studies of specific biochemical or genetic events. Here, we briefly describe four types of bacteriophages, all of which infect *E. coli*, that have been especially useful in molecular biology research.

DNA PHAGES OF THE T SERIES

The T phages of E. coli are large lytic phages that contain a single molecule of double-stranded DNA. This molecule is about 2×10^5 base pairs long in T2, T4, and T6 viruses and about 4×10^4 base pairs long in T1, T3, T5, and T7 viruses. T-phage virions consist of a helical protein "tail" attached to an icosahedral "head" filled with the viral DNA. After the tip of a T-phage tail adsorbs to receptors on the surface of an E. coli cell, the DNA in the head enters the cell through the tail. The phage DNA then directs a programme of events that produces approximately 100 new phage particles in about 20 minutes, at which time the infected cell lyses and releases the new phages.

The initial discovery of the role of messenger RNA in protein synthesis was based on studies of E. coli cells infected with bacteriophage T2. By 20 minutes after infection, infected cells synthesize T2 proteins only. The finding that the RNA synthesized at this time had the same base composition as T2 DNA (not E. coli DNA) implied that mRNA copies of T2 DNA were synthesized and used to direct cellular ribosomes to synthesize T2 proteins.

TEMPERATE PHAGES

Bacteriophage λ, which infects E. coli, typifies the temperate phages. This phage has one of the most studied genomes and is used extensively in DNA cloning. On entering an E. coli cell, the double-stranded λ DNA assumes a circular form, which can enter either the lytic cycle (as T phages do) or the lysogenic cycle. In the latter case, proteins expressed from the viral DNA bind a specific sequence on the circular viral DNA to a similar specific sequence on the circular bacterial DNA. The viral proteins then break both circular molecules of DNA and rejoin the broken ends, so that the viral DNA becomes inserted into the host DNA. The carefully controlled action of viral genes maintains λ DNA as part of the host chromosome by repressing the lytic functions of the phage. Under appropriate stimulation, the λ prophage is activated and undergoes lytic replication.

SMALL DNA PHAGES

The genome of some bacteriophages encodes only 10 – 12 proteins, roughly 5 – 10 per cent of the number encoded by T phages. These small DNA phages are typified by the Ö×174 and the filamentous M13 phages. These were the first organisms in which the entire DNA sequence of a genome was determined, permitting extensive understanding of the viral life cycle.

The viruses in this group are so simple that they do not encode most of the proteins required for replication of their DNA but depend on cellular proteins for this purpose. For this reason, they have been particularly useful in identifying and analyzing the cellular proteins involved in DNA replication.

RNA PHAGES

Some E. coli bacteriophages contain a genome composed of RNA instead of DNA. Because they are easy to grow in large amounts and because their RNA genomes also serve as their mRNA, these phages are a ready source of a pure species of mRNA. In one of the earliest demonstrations that cell-free protein synthesis can be mediated by mRNA, RNA from these phages was shown to direct the synthesis of viral coat protein when added to an extract of E. coli cells containing all the other components needed for protein synthesis. Also, the first long mRNA molecule to be sequenced was the genome of an RNA phage. These viruses, among the smallest known, encode only four proteins: an RNA polymerase for replication of the viral RNA, two capsid proteins, and an enzyme that dissolves the bacterial cell wall and allows release of the intracellular virus particles into the medium.

ANIMAL VIRUSES

Animal viruses come in a variety of shapes, sizes, and genetic strategies. In this book, we are concerned with viruses that exhibit at least one of two features: they utilize important cellular pathways to form their molecules, thereby closely mimicking a normal cellular function, or they can integrate their genomes into those of normal cells.

The names of many viruses are based on the names of the diseases they cause or of the animals or plants they infect. Common examples include poliovirus, which causes poliomyelitis; tobacco mosaic virus, which causes a mottling disease of tobacco leaves; and human immunodeficiency virus (HIV), which causes acquired immunodeficiency syndrome (AIDS).

However, many different kinds of viruses often produce the same symptoms or the same apparent disease states; for example, several dozen different viruses can cause the red eyes, runny nose, and sneezing referred to as the common cold. Clearly, any attempt to classify viruses on the basis of the symptoms they produce or their hosts obscures many important differences in their structures and life cycles.

What *are* central to the life cycle of a virus are the types of nucleic acids formed during its replication and the pathway by which mRNA is produced. The relation between the viral mRNA and the nucleic acid of the infectious particle is the basis of a simple means of classifying viruses. In this system, a viral mRNA is designated as a *plus strand* and its complementary sequence, which cannot function as an mRNA, is a *minus strand*.

A strand of DNA complementary to a viral mRNA is also a minus strand. Production of a plus strand of mRNA requires that a minus strand of RNA or DNA be used as a template. Using this system, six classes of animal viruses are recognized. Bacteriophages and plant viruses also can be classified in this way, but the system has been used most widely in animal virology because representatives of all six classes have been identified.

Bibliography

Alon, U.: *An Introduction to Systems Biology: Design Principles of Biological Circuits*, London: Academic Press, 2006.

Andrewartha, H.: *The Distribution and Abundance of Animals*, Chicago: Illinois, University of Chicago Press, 1998.

Ashby, R.: *Principles of the Self-organizing Dynamic System*, New Jersey: Princeton University Press, 2004.

Attenborough, D.: *Life on Earth, a Natural History*, London, BBC/ Collins, 1999.

Baker, J.: *The Study of Biology*, New York: Columbia University Press, 1997.

Banathy, B.: *A Systems View of Education; Concepts and Principles for Effective Practice*, New Jersey: Educational Techology Publications, 2002.

Bardach, A.: *The Chemical Senses of Fishes*, London: Academic Press, 2005.

Barnett, L.: *The World we Live*, New Jersey: Princeton University Press, 2001.

Bock, G.: *Simulation of Biological Processes*, London: Academic Press, 2000.

Bruggeman, F.: *Systems Biology: Philosophical Foundations*, New York: Pergamon, 2008.

Ellner, SP.: *Dynamic Models in Biology*, Princeton: Princeton University Press, 1998.

Flood, R.: *An Introduction to the Theory and Application of Systems Science*, New York: Plenum Press, 1999.

Herwig, R.: *Systems Biology in Practice,* New York: Columbia University Press, 2005.

Kaneko, K.: *An Introduction to Complex Systems Biology,* New York: John Wiley and Sons, 2001.

Kitano, H.: *Foundations of Systems Biology,* New York: New American Library, 2003.

Klir, G.: *Facets of Systems Science,* Plenum Press: New York, 2007.

Miller, J.: *Living Systems,* New York: McGraw Hill Book Company, 2003.

Palsson, B.: *Systems Biology-properties of Reconstructed Networks,* Cambridge: Cambridge University Press, 2006.

Rapoport, A.: *The International Encyclopedia of Social Sciences,* New York: Macmillan, 2008.

Schneider, E. *Thermodynamics, and Life,* Chicago: The University of Chicago Press, 2000.

Skyttner, L.: *General Systems Theory: Problems, Perspectives,* New Jersey: World Scientific, 2002.

Index

D

E

G

H

I

M

N

O

P

R

S

T

V

W

Z